Springer INdAM Series

Volume 6

For further volumes:
http://www.springer.com/series/10283

Alessandra Celletti · Ugo Locatelli ·
Tommaso Ruggeri · Elisabetta Strickland
Editors

Mathematical Models and Methods for Planet Earth

Springer

Editors
Alessandra Celletti
Dipartimento di Matematica
Università degli Studi di Roma
"Tor Vergata"
Roma, Italy

Tommaso Ruggeri
Dipartimento di Matematica
Università degli Studi di Bologna
Bologna, Italy

Ugo Locatelli
Dipartimento di Matematica
Università degli Studi di Roma
"Tor Vergata"
Roma, Italy

Elisabetta Strickland
Dipartimento di Matematica
Università degli Studi di Roma
"Tor Vergata"
Roma, Italy

ISSN: 2281-518X ISSN: 2281-5198 (electronic)
Springer INdAM Series
ISBN 978-3-319-35004-2 ISBN 978-3-319-02657-2 (eBook)
DOI 10.1007/978-3-319-02657-2
Springer Cham Heidelberg New York Dordrecht London

Cover Design: Raffaela Colombo, Giochi di Grafica, Milano, Italy
Typesetting with LaTeX: PTP-Berlin, Protago TeX-Production GmbH, Germany (www.ptp-berlin.de)

Springer is part of Springer Science+Business Media (www.springer.com)

Preface

More than a hundred scientific societies, universities, research institutes and organizations from around the world have banded together to dedicate 2013 as a special year for the Mathematics of Planet Earth. The Italian partner which endorsed the MPE2013 initiative is INdAM, the National Institute of Advanced Mathematics. First founded in 1939 by the noted mathematician Francesco Severi, the aims of INdAM are the training of researchers in mathematics, especially in the emerging branches, in order to foster the transfer of knowledge to technological applications and promote contact between Italian and international mathematical research. In order to achieve its objectives in training researchers and in supporting excellence programs, INdAM offers fellowships from undergraduate level to experienced researchers, as well as organizing workshops, meetings and schools.

INdAM has decided to share in and support the mission of the MPE2013, as it shares the desire of the world's mathematical community to learn more about the challenges faced by our planet and the underlying mathematical problems. The MPE2013 initiative organized by INdAM took place in Rome, Italy, on May 27–29, 2013, and consisted in the Workshop "Mathematical models and methods for Planet Earth", organized by Alessandra Celletti (Università di Roma Tor Vergata), Ugo Locatelli (Università di Roma Tor Vergata), Tommaso Ruggeri (Università di Bologna) and Elisabetta Strickland (Università di Roma Tor Vergata).

Over the course of the workshop an international group of mathematicians (namely all corresponding authors of the following chapters and other outstanding invited speakers) with a very wide range of expertise in various branches presented findings on several themes related to the MPE2013: Earth as a planet to discover, a planet supporting life, a planet organized by humans, and a planet at risk from celestial threats. The topics of the talks concerned social, biological, medical, geological and astronomical problems related to our planet. Mathematical methods for studying complex systems arising in the fields of social prevention and socio-economic problems were presented with the aim, for example, to understand emerging collective behaviours of a high number of interacting units. In particular, stochastic models can be used to investigate complex social and biological behaviours. In biology and medicine, mathematics plays a pivotal role through modelling and simulations in a diverse range of contexts: from the behaviour of cells and tissues to the description of tumour growths. The leading role of mathematics in supporting our planet is also witnessed by the authorship attribution of literature texts and by models for future internet information dissemination.

Beyond the investigation of human-related aspects, mathematics allows us to study the physical characteristics of our planet. Most notably, some talks were devoted to the calibration of the geological time scales (a crucial aspect which allows us to retrieve specific events in Earth history), to the investigation of boundary layers associated with large-scale ocean circulation and, farther up in the atmosphere, to studying the Earth's climate variability and changes using the theory of dynamical systems.

Safeguarding the Earth is not limited to our planet and its atmosphere but - as we are only one part of the solar system - one really needs to investigate the interaction of the Earth with the other bodies populating the neighbouring sky. Indeed, the investigation of the so-called N-body problem allows us to study the stability of the Earth's dynamics as well as to identify new interplanetary trajectories. The recent impact of the meteorite in Chelyabinsk (Russia) alerted mankind to the necessity of protecting the planet from near-Earth asteroid hazards and of developing mitigation strategies. Finally, we are also definitely worried by the thousands of pieces of space debris from defunct satellites and fragments, which now surround the Earth and form a dangerous envelope: a mathematical investigation of the dynamics of space debris has now become vital.

In addition to the talks devoted to the investigation of the above topics, a special event of the Workshop was the public lecture by Christiane Rousseau (Université de Montréal), vice-president of the International Mathematical Union. Her talk dealt with the complexity of the Earth as a whole and outlined the role of mathematics in protecting and discovering our planet.

One interesting aspect that came out of this meeting is that the amount of data involved in some scientific problems has become overwhelmingly large, so that there is an apparent loss of simplicity between mathematics and its applications. Today no one could so deeply master as many mathematical arguments as Poincaré or Hilbert were able to do a century ago; but all the speakers at the workshop clearly showed that mathematicians are now also challenged in the opposite direction: more and more research topics require deep mathematical knowledge, often to tackle problems in the context of network teams. Though abstraction can allow mathematics to remain pure, mathematicians are called upon to more intensively work together with the rest of the scientific community.

INdAM believes that the collaborations and efforts of all scientists who participated in the workshop pointed out that our planet is home to dynamic processes of all sorts. The challenges facing our planet and our civilization are multidisciplinary and multifaceted, and the mathematical sciences play a central role in scientific efforts to understand and effectively address those challenges. INdAM sincerely believes that MPE2013 will also help us to motivate students and young researches by providing stimulating answers to questions like: "What is mathematics good for?". We conclude by quoting Marta Sanz-Solé, President of the European Mathematical

Society; at the UNESCO Headquarters in Paris during the MPE Day on 5 March 2013, she stated that "The MPE2013 initiative will expose mathematicians to the whole world, by showing their usefulness and stimulating research. From now on, mathematics can no longer be associated with a pure intellectual exercise without connection to the most important problems of mankind". We hope that the INdAM Workshop has contributed to pursuing those goals.

Roma, 2013

Alessandra Celletti
Ugo Locatelli
Tommaso Ruggeri
Elisabetta Strickland

Contents

Mathematics of Planet Earth

Christiane Rousseau

Abstract Why did I start MPE2013? Maybe because I am a dreamer... I love nature, I love mathematics and I am concerned with the planet. My own research is not in applied mathematics, but I like popularizing mathematics and when I do so, I usually present themes that are far from my own research. Indeed, one of my role as scientist is to "digest" research papers, to extract the ideas and to explain then in simple terms how mathematics is developing within science around us. From the beginning, the theme of Mathematics of Planet Earth appeared to me as very rich. Now, four years after I started MPE2013, the theme appears as immense and one of my rewards for coordinating MPE2013 is to continuously learn new MPE topics. At the same time, I am becoming more militant for the planet. In this note, I will describe to you some of facets of MPE that I have discovered with the hope that you will yourself get interested in the mathematics behind discovering, understanding, organizing and protecting our planet.

To help exploring "Mathematics of Planet Earth" we often divide it into four themes:

- A planet to discover.
- A biologically diverse planet.
- A planet organized by civilization.
- A planet in danger.

1 A planet to discover

One of my favorite theme when I do outreach is that mathematics allows to transform data and observations into some vision of an object: we put our "mathematical

C. Rousseau (✉)
Département de mathématiques et de statistique, Université de Montréal, C.P. 6128, succ. Centre-ville, Montréal, Qc, H3C 3J7, Canada
e-mail: rousseac@dms.umontreal.ca

A. Celletti, U. Locatelli, T. Ruggeri and E. Strickland (eds.): *Mathematical Models and Methods for Planet Earth.* Springer INdAM Series 6, DOI 10.1007/978-3-319-02657-2_1,

glasses" and we "see" the Earth. This was done all over the centuries. Already the Greeks knew that the Earth was a sphere and Eratosthenes calculated the approximate circumference of the Earth by remarking that while the Sun was vertical at Syene it would make an angle of 7.5^o with the vertical direction at Alexandria.

From Newton's law of universal gravitation $F = \frac{GmM}{r^2}$, and knowing that the gravitational acceleration at the surface of the Earth is approximately $g = 9.8\,\mathrm{m/s^2}$, that $g = \frac{GM}{r^2}$, as well as the values of G and of the radius R of the Earth, we can calculate the Earth's mass: $M \simeq 5.98 \times 10^{24}$ kg. The Earth is much too heavy to be homogeneous since the density of the crust is around 2.2–$2.9\,\mathrm{kg/dm^3}$ and the mean density of $5.52\,\mathrm{kg/dm^3}$. This means that the interior of the Earth is very heavy!

Exploring the interior of the Earth can be done through analysis on the seismic waves generated by large earthquakes. Richard Dixon Oldham identified the different types of seismic waves recorded on seismographs. These include the pressure waves (or P-waves) which travel both through the solid and viscous layers, while the shear waves (or S-waves) are stopped by the viscous layers. This allows identifying the viscous layers of the Earth. The Danish mathematician, Inge Lehmann, went further. In 1936, she discovered the inner core of the Earth, which is now admitted to be solid. She was working at the Danish Geodetic Institute and analyzed the measures of the different travel times of seismic waves generated by major earthquakes to different stations over the Earth. If the Earth were uniform, then the signal would travel along straight lines with travel time as in Figure 1. At the time, it was known that there was a discontinuity between the mantle and the core, and that the waves were traveling slower in the core, leading to the pattern of Figure 2 because of the refraction of the waves when entering the core. No wave is detected for center angle in $[112^o, 154^o]$. But Inge Lehamnn discovered that some waves were indeed recorded in the forbidden region. A piece was missing for the puzzle... This missing piece was the inner core, in which the waves travel faster, thus allowing that the waves arriving tangentially to the inner core be reflected on it (see Fig. 3).

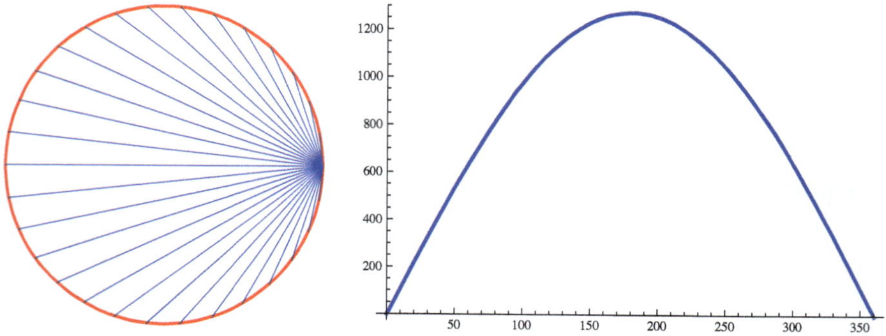

Fig. 1 The paths and travel times of the seismic waves in a homogeneous Earth for a speed of 10 km/s and a radius of the Earth of 6360 km. Left panel: the paths of the waves. Right panel: the travel time (in seconds) of the waves reported in the left panel depending on the angular coordinate (in degrees) of the end point of the wave

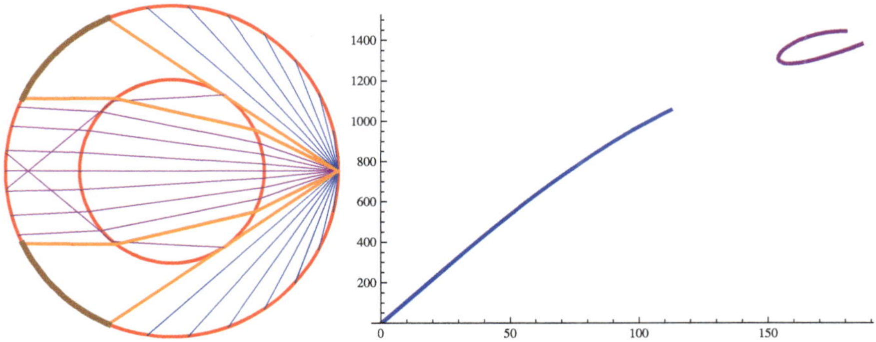

Fig. 2 The paths and travel times of the seismic waves in the Earth for a speed of 10 km/s inside the mantle and of 8 km/h inside the core. Left panel: the paths of the waves. Right panel: the travel time (in seconds) of the waves reported in the left panel depending on the angular coordinate (in degrees) of the end point of the wave. Note that refracted waves can intersect each other, thus explaining the shape of the curve of travel times

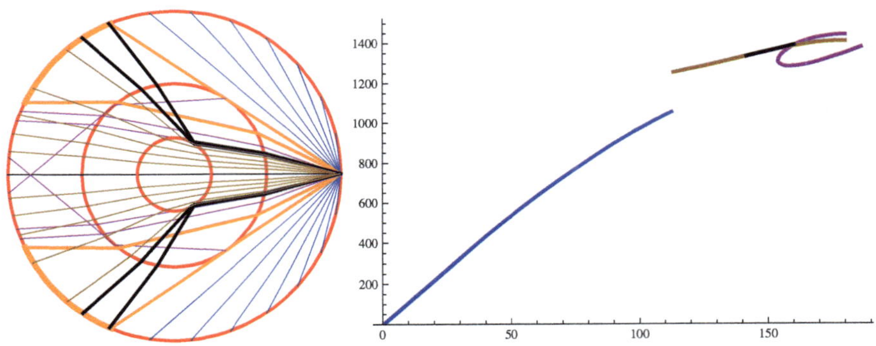

Fig. 3 The paths and travel times of the seismic waves in the Earth for a speed of 10 km/s inside the mantle, 8 km/h inside the outer core, and 8.8 km/h inside the inner core. Left panel: the paths of the waves: the paths between the two black lines are reflected on the inner core, while the others are only refracted. Right panel: the travel time (in seconds) of the waves reported in the left panel depending on the angular coordinate (in degrees) of the end point of the wave

More recently, remote sensing methods allow localizing petrol or gas in the Earth by analyzing reflected signals sent by air-guns. But only the signals of large earthquakes are sufficiently powerful to analyze finer details at greater depths. These signals have been analyzed by Raffaella Montelli to provide evidence for the conjecture that the isolated volcanic islands, like the Hawaiian Islands, were produced by volcanic plumes though the mantle (Fig. 4). The movement of the tectonic plates would then explain that the islands are aligned and ordered in terms of increasing age. Wavelets are particularly suited for analyzing such fine details and the analysis was done recently by Ingrid Daubechies together with the geophysicists Tony Dahlen and Guust Nolet.

Describing the Earth also means many other things, in particular describing the convection movements in the mantle, and the Earth's oceans and climates. But we

Fig. 4 The creation of volcanic islands by a volcanic plume through the mantle while the upper tectonic plate is moving. Left panel: a volcanic plume through the mantle. Right panel: islands have been created at a later stage. (The figures are taken from Wikipedia, http://en.wikipedia.org/wiki/File:Hotspot%28geology%29-1.svg)

will limit ourselves to just another fascinating and very modern subject: the planetary motions of the Earth.

The planetary motions of the Earth. Lagrange was the first to have had the idea that the past climates of the Earth including the glaciation periods could be explained through variations of the parameters of the elliptical orbit of the Earth around the Sun: the periodic oscillations of the major axis, of the eccentricity, of the orientation and inclination of the Earth axis with respect to the elliptic plane are called the Milankovitch cycles. But there is no proof that these oscillations will remain for ever and, indeed, Jacques Laskar showed in 1989 that the inner planets have chaotic orbits and that we could not exclude that the orbit of Venus could cross that of Mercury. What characterizes chaos is sensitivity to initial conditions, which means that any (unavoidable) errors that we make on the initial conditions lead to very large errors after several millions years. While any simulated past or future trajectory of the Earth may be completely different from the real trajectory of the Earth, the shadowing lemma ensures that such a trajectory is a realistic trajectory for a planet that would be just a bit different from the Earth, or at a slightly different initial position. Hence, making many simulations in parallel describes potential futures for the planets with associated probabilities. This is what Jacques Laskar achieved in 2009 after having simulated 2500 scenarios in parallel. Several of these scenarios show crossing of the orbits of the inner planets, Mercury, Venus, the Earth, and Mars, which could lead to collisions or to the ejection of planet(s) from the solar system.

Since the inner planets have chaotic motions, the obliquity of their axis could have very large oscillations. This has been the case with Venus in the past and is presently the case for Mars. However, the obliquity of the Earth's axis has only very small oscillations with amplitude 1.2°. Why? Jacques Laskar showed in 1993 that the Moon protects us: without the Moon, the Earth's axis would have very large oscillations, similar to those of Mars' and Venus' axis.

2 A biologically diverse planet

Earth is inhabited by millions of species. Where does all this biodiversity come from? Mutations (randomness) create new species. These species interact to survive, and it is a challenge for the scientists to understand how the biodiversity maintains itself through the appearance and extinction of new species. Hence, it is natural to introduce models for interactions of species. These models are best represented geometrically through the phase portrait which gives a geometric description of the evolution of the populations. This is the case of the simple Lotka-Volterra model for predator-prey interaction which yields periodic oscillations of the populations of predator and prey, the amplitude of the oscillations and the length of the period depending on the initial conditions (see Fig. 5). Similar models can be given for two competing species. As in the case of predator-prey interaction, the system depends on parameters, and here we can obtain four different qualitative behaviors depending on the parameter values. They are represented in Figure 6. There, we can see that, as soon as the competition is sufficiently strong, one species disappears.

This has been generalized by Simon Levin in the case of more than 2 species. The model for n species $x_1,\dots,x_n$, takes the form

$$\frac{dx_i}{dt} = x_i f_i(x_1,\dots,x_n,y_1,\dots y_m), \qquad i = 1,\dots,n,$$

where the y_i could be m resources or other regulating factors that are not significantly affected by the species x_j, and the functions f_i are affine. It is possible to identify a minimal set of regulating factors $\{z_1,\dots,z_k\}$, where $k \leq m+n$, the z_j are affine functions of $x_1,\dots,x_n,y_1,\dots y_m$, and the functions f_i depend on $z_1,\dots,z_k$ alone. The result of Levin is that if $n > k$, then no more than k species will survive. It hence

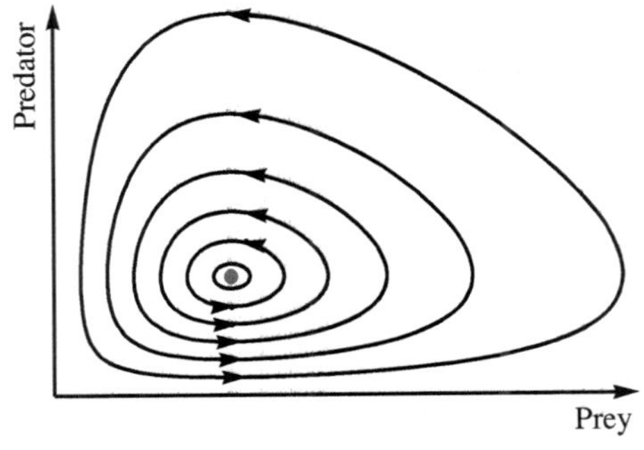

Fig. 5 Phase portrait for Lotka-Volterra predator-prey system. If we start with small initial conditions of predator and prey, then the prey increases. When there is more prey, then the predator increases. When the population of predator becomes sufficiently large, this yields a decrease of prey. The predator then decreases when the food is becoming rare and the cycle starts again

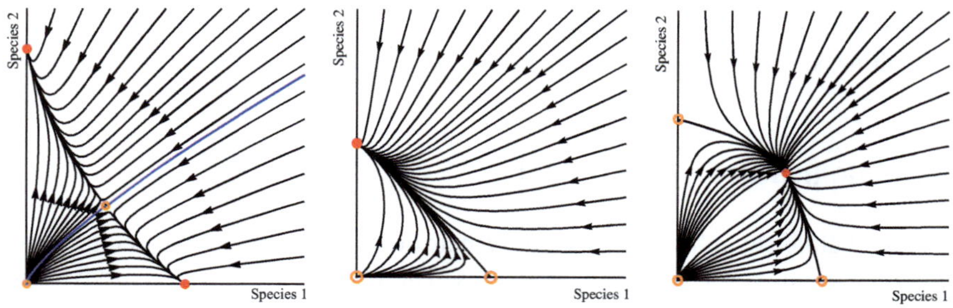

Fig. 6 The case in the left panel represents strong competition and leads to the extinction of one species depending on the initial condition. The case in the mid panel leads to the extinction of species 1; the symmetric case exists with the extinction of species 2. The weak competition of the case in the right panel allows the coexistence of the species

seems that competition goes against biodiversity! So, we need other forces to maintain biodiversity.

One of them is *spatial heterogeneity*. Indeed, suppose that we have strong competition between two species as in the left box of Figure 6. It could happen that in some regions we have initial conditions leading to the survival of species 1, and in other regions we have initial conditions leading to the survival of species 2. A second force helping maintaining biodiversity is *temporal heterogeneity*. The parameters of the system may change over time, so we may switch over time between the four different cases (the three cases of Fig. 6 and the symmetric case of the mid box case). For instance, one species may be more resistant to drought periods, while the other is more resistant to winter, etc.

Another force, *cooperation*, supports biodiversity. The principle is best explained through the prisoner's dilemma (see Table 1). From the Individual 1 point of view, if he supposes that Individual 2 chooses at random his strategy, then he expects to spend a mean of 2 years in jail if he defects and a mean of 3 years in jail if he cooperates. Hence, his optimal choice is to defect. But, why should Individual 1 make the hypothesis that Individual 2 chooses his strategy at random? Most likely, Individual 2 will make the same reasoning as Individual 1 and will also choose to defect. Then they will both have three years in jail. They could have had only two if they had pushed the reasoning one step further and decided that it was in their common interest to cooperate.

Table 1 The prisoner's dilemma

Individual 1 \ *Individual 2*	*Cooperate (remain silent)*	*Defect (confess)*
Cooperate (remain silent)	**2 years in jail** 2 years in jail	**4 years in jail** 1 year in jail
Defect (confess)	**1 year in jail** 4 years in jail	**3 years in jail** 3 years in jail

Martin Nowak made computer experiments on the evolution of populations of defectors and cooperators. He started with random distribution of defectors and cooperators. At each round, the winners produced offspring who participate in the next round. Within a few generations, all individuals were defecting. Then he identified five mechanisms of cooperation: introducing them in the game led to communities dominated by cooperators. These five mechanisms are observed in nature and/or in the human behaviour:

1. **Direct reciprocity:** this mechanism occurs when there are repeated encounters between the same individuals who learn from their previous choices. For instance, vampire bats share with the bat who found no blood. A winning strategy in this case could be tit-for-tat, namely starting by cooperating and then do the last move of the other player. A more robust strategy in case of random mistakes is win-stay, lose-shift.
2. **Network selection:** this occurs when cooperators and defectors are not uniformly spatially distributed, and individuals interact more often with their neighbors. This leads to patches of cooperators and patches of defectors. Such a mechanism occurs for instance with yeast cells.
3. **Kin selection:** cooperation (including sacrifice) between genetically related individuals.
4. **Indirect reciprocity:** help of another based on the needy's individual reputation, for instance with Japanese macaques.
5. **Group selection:** employees competing among themselves, but cooperating for their company.

Cooperation has modeled the world as we know it. It explains the preservation of biodiversity. It is everywhere present in the human organization of the planet.

3 The planetary challenges

The problems are very complex since all systems are intertwined. BUT, already in the Stern Review on the Economics of Climate Change in 2006, we can read that *the benefits of strong, early action on climate change far outweigh the costs for not acting*. Hence, we should convince governments to act. But how? We have to pay now and benefits will only be felt in the period between 50 years and 200 years from now!

Also, the problems will be irregularly felt around the world. Some countries might benefit from the climate warming (new areas opening to agriculture, trees growing faster), while others will be destroyed or ruined. And it is not necessarily the same countries that are contributing to the increase of green house gas and that will suffer from the consequences. Here again, we find the prisoner's dilemma in another form: the dilemma of depolluting (see Table 2). Since green-house gas spread over the whole planet, all countries suffer from the pollution done by the delinquent countries, and hence, endure costs for that. And the delinquent countries benefit from the efforts of the others!

Table 2 The dilemma of depolluting. It costs two units per country for each country polluting and it costs 3 units for a country to depollute. (These numbers have of course to be adjusted to the real costs)

Country 1 \ *Country 2*	*Depolluting*	*Polluting*
Depolluting	**Cost : 3** Cost : 3	**Cost : 5** Cost : 2
Polluting	**Cost : 2** Cost : 5	**Cost : 4** Cost : 4

The real consequences are unknown. We could expect ecosystems to disappear and be replaced by others. How will the transition take place? Smoothly with new species of plants and animals installing themselves among the old ones? Or abruptly with all trees dying and an intermediate period with no forest before new forests develop? For instance, in the Western part of North America the mountain pine beetle is destroying large territories of forests. This insect has always existed. Its population would decrease drastically during the winters, thus allowing the population to remain under control. Recently, the milder winters have allowed the population to grow enormously.

A general feature which applies to all these problems is that **a model only contains what we put in it**. The idea of modeling is to put in a model exactly the significant ingredients and to leave the other ones aside, so as to highlight the main features and tendencies. But what if we forget essential ingredients? Let's take the example of the Arctic warming. Less than 10 years ago we would hear that the Arctic Ocean would be free of ice during the summer by 2050. Now, we expect it to be free of ice much sooner, maybe even before 2020. What has happened? If we look at a graph of the average monthly Arctic sea ice extent for 33 years as in the left box of Figure 7, then it is tempting to pass a regression line which fits quite well with the data. The regression line always exists, but are we allowed to interpolate from it on longer periods?

Let's look at the right box of Figure 7: there, the regression line predicts that the Arctic Ocean could be free of ice by 2034, while in case of the left box the predic-

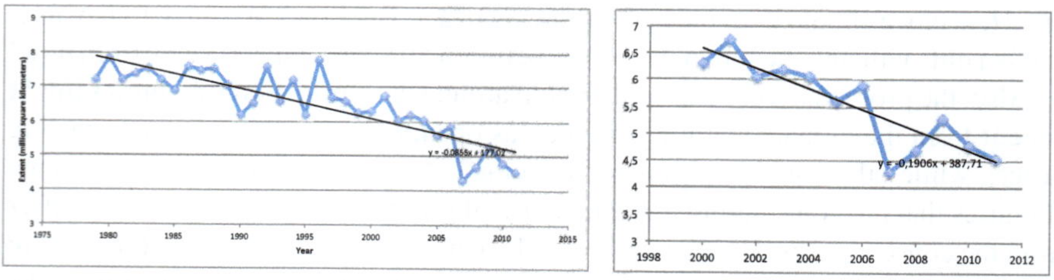

Fig. 7 Average monthly sea ice extent from 1979 to 2011. Graphic drawn using data from National Snow and Ice Data Center (US). Left panel refers to the years from 1979 to 2011; the regression line predicts a zero average sea ice extent by 2070. Right panel refers to the years from 2000 to 2011; in this case the line predicts a zero average by 2034

tion is for 2070. Do we have a good reason to think that the extent of Arctic sea ice depends linearly on time? If yes, then we should choose the graphic on the left since it contains more data. But we could also question if extrapolating with a regression line is allowed. Indeed, we have to take into account the albedo effect, namely the fact that the Sun heat is reflected on ice while absorbed by the ocean when the ice is melt, thus starting a feed-back loop which intensifies the phenomenon: this is not a linear phenomenon. We now realize that methane, a very strong green house gas, is released by the oceans. And the permafrost is melting in Northern regions and releases large quantities of methane. All these ingredients have to be added to the models. **Have we forgotten other essential ingredients?** This illustrates the difficulty of putting exactly the significant ingredients in a model and the importance of interdisciplinary efforts to advance research on these problems.

Several countries have understood that we should move to sustainable development. But what means sustainable development? This was defined by the Bruntland Commission of the United Nations in 1987: *It is development that meets the needs of the present without compromising the needs of the future.* Is this definition sufficiently precise to guide our actions? Let's look at an example.

The case of fisheries. Let us suppose we have models for the evolution of populations of fish that are harvested. Fishermen go fishing if they get more money for their catches than what they spend to go fishing. Otherwise, they stop fishing. Hence, with free access to the resource and with prices increasing when the resource is rare, then we can run short of fish. So let us control the access. How? The answer seems obvious. *The quantity that fishermen are allowed to catch is chosen so as to maximize the revenue obtained from the resource over the years.*

The answer is not so obvious ... Colin Clark showed in 1973 that if the discount factor is at least twice the reproduction rate, then we maximize the revenue by fishing the whole population now and putting the money in the bank! This is the case for populations of fish that reproduce very slowly (deep sea fish). This example illustrates well the difficulty of decision making.

As a conclusion, there are more questions than answers. We discover new pressing problems faster than the known ones are solved. For instance, we could believe a few years ago that the price of oil would increase significantly with the increasing demand, and that this would encourage the development of green technologies, even if this development requires significant financial investments. Very recently, the discovery of massive reserves of shale oil and gas has changed the rules of the game and may affect the cost-effectiveness of certain investments in green energy.

Mathematics has an essential role to play in these issues. This is why MPE2013 was necessary. And we should hope that the interest for MPE and the increased collaboration between the scientists and researchers in mathematical sciences it created does not stop at the end of 2013.

References

1. Official web site of "Mathematics for Planet Earth", http://www.mpe2013.org/

The Role of Boundary Layers in the Large-scale Ocean Circulation

Laure Saint-Raymond

Abstract Understanding the mechanisms governing the ocean circulation is a challenge for geophysicists, but also for mathematicians who have to develop tools to analyze these complex models (involving a large number of time and space scales).

A particularly important mechanism for the large-scale circulation is the boundary layer phenomenon, which accounts for a macroscopic part of the energetic fluxes. We will show here using a very simplified model that it explains in particular the Western intensification of currents. We will then exhibit the mathematical difficulties arising in more complex geometries.

1 Introduction

A precise description of the oceanic circulation is fundamental both for the understanding of sea life (salinity and temperature, which are carried by the current, determine plant and animal life), and for human activities such as shipping and fishing. Our goal here is to understand the Westward Intensification of Wind-Driven Ocean Currents (see Fig. 1).

The theoretical approach we will describe does not aim at providing quantitative predictions. It rather consists in proposing qualitative processes for the formation and propagation of such structures. For that purpose, we will work with a toy model, or in other words with the simplest possible model which exhibits the expected phenomenology, creating western intensification as response to the wind forcing.

L. Saint-Raymond (✉)
Département de Mathématiques et Applications, Université Pierre et Marie Curie & Ecole Normale Supérieure, 45 rue d'Ulm, 75230 Paris Cedex, France
e-mail: Laure.Saint-Raymond@ens.fr

A. Celletti, U. Locatelli, T. Ruggeri and E. Strickland (eds.): *Mathematical Models and Methods for Planet Earth.* Springer INdAM Series 6, DOI 10.1007/978-3-319-02657-2_2,

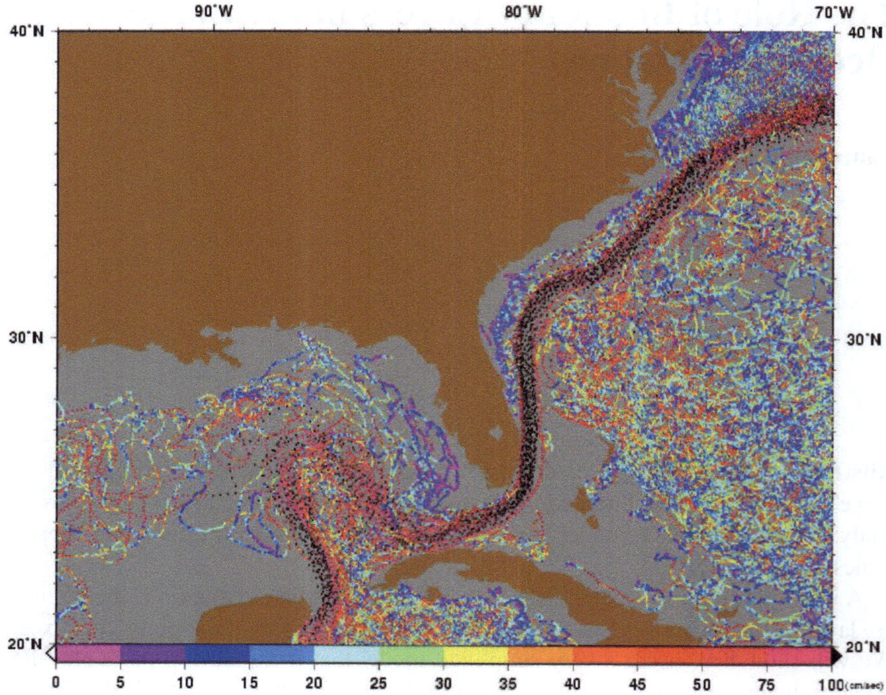

Fig. 1 The Westward intensification of oceanic currents along Florida. Reproduced with permission from [17]

2 A two-dimensional mathematical model: the Munk equation

We therefore focus on the main competing mechanisms. We start from the assumption that the phenomenon is essentially dynamical, and that we can neglect the effects due to the variations of density, temperature and salinity of water, as well as the vertical structure of the circulation (the ocean depth being very small compared to its horizontal extent). The only unknown is thus the local (two-dimensional) velocity $(u_1, u_2) \equiv (u_1(x_1, x_2), u_2(x_1, x_2))$ where (x_1, x_2) denote respectively the longitude and latitude.

2.1 Seawater, an incompressible and weakly viscous fluid

- Seawater is essentially incompressible, and homogeneous. Physical observations indeed show that the variations of the density are essentially negligible outside from a thin surface layer, referred to as the pycnocline.
  These small vertical variations also account for the fact that – in first approximation – we can forget the stratification and deal with 2D models.

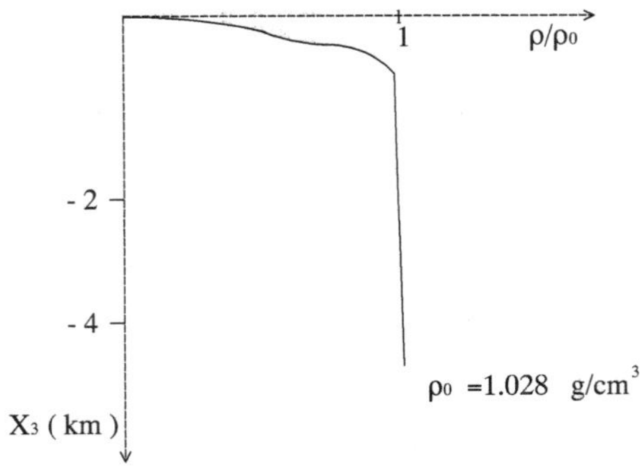

Fig. 2 Density profile depending on the depth

The equation for the conservation of mass then states

$$\partial_1(\rho_0 u_1) + \partial_2(\rho_0 u_2) = 0,$$

where ρ_0 is the (constant) density of water.

- The other physical feature of seawater to be discussed is its viscosity. We first notice that the kinematic viscosity of water is negligible, and does not account for the energy dissipation. The energy input due to wind forcing and solar heating has indeed to be compensated by some dissipation mechanism. A classical way to catch this phenomenon is to introduce a turbulent viscosity, which is supposed to model the effect of small scales.

  Note however that there is even no heuristic argument to justify such a model.

2.2 *The Coriolis force and the role of wind*

Seawater is submitted to different forces. As mentioned at the beginning, we will neglect thermodynamic effects due to temperature variations, although they are known to be one of the main contribution to the general circulation, also called thermohaline circulation. This is one of the main drawback of our simplified model, and future works will have clearly to deal with this effect.

- The Coriolis force takes into account the Earth rotation: we indeed focus on the departure of the fluid motion from the solid-body rotation, meaning that we consider a non Galilean reference frame.
  In the general 3D framework, a simple change of variables leads to the following expression

  $$F_{Coriolis} = 2\rho_0 \omega \wedge u \text{ where } \omega \text{ is the (constant) rotation vector.}$$

  In bidimensional models, the main horizontal contribution at latitude θ_0, called f-plane approximation, is

  $$fu^\perp(x_1,x_2) \text{ with } f = 2\omega\cos\theta_0$$

denoting by $u^\perp$ the vector field of coordinates $(u_2, -u_1)$. Note that it modifies only the pressure.
The next contribution, referred to as β-plane approximation, is due to inhomogeneities

$$\beta x_2 u^\perp(x_1, x_2) \text{ with } \beta = \frac{2\omega \sin\theta_0}{L}$$

where L is the typical horizontal length.

- On the other hand, experimental observations show that currents are strongly correlated to the wind (see http://www.aviso.oceanobs.com/en/applications/ocean/large-scale-circulation/currents-around-the-world/monsoon-currents.html).
An analytical computation of the forcing was proposed by Ekman. It explains how the surface stress due to the wind can be transferred to the whole mass of water : this mechanism, which is actually very similar to the boundary layer phenomenon we will study here, is referred to as Ekman's pumping.

Note however that a more realistic model should take into account a real coupling with the atmosphere. We indeed expect the wind not to be prescribed independently of the ocean currents.

2.3 A balance equation

For a stationary flow, the acceleration which is the sum of the geostrophic focus (Coriolis force and pressure gradient), viscous dissipation $-\nu\Delta u$ and wind forcing σ, vanishes. The equation states

$$\beta x_2 u^\perp + \nabla p - \nu \Delta u + \sigma = 0 \text{ in } \Omega. \tag{1}$$

Note that the pressure p is determined as the Lagrange multiplier associated to the incompressibility constraint.

This system of partial differential equations of order 2 is supplemented by some boundary condition. The no-slip condition, also called Dirichlet boundary condition, states

$$u_{|\partial\Omega} = 0.$$

Multiplying by u and integrating by parts, we find that

$$\nu \int |\nabla u|^2 dx + \int \sigma \cdot u dx = 0,$$

meaning that the energy input due to the wind is exactly balanced by the viscous dissipation.

3 The boundary layer phenomenon: a recent discovery

At this stage, equations are so simplified that one could almost compute – analytically or numerically – their solutions using for instance spectral decompositions. Nevertheless, because of the different scales arising in the problem, the formulæ we obtain in such a way are too complex to allow a rapid envisioning of the geometry of the flow.

3.1 The pionneering work of Prandtl

A natural idea is to decompose the flow as a superposition of elementary contributions with different scalings, that is to introduce some multi scale expansion of the solution to the Munk equation (1).

The first attempt to implement this strategy for singular boundary problems goes back to Prandtl in his pioneering work [20]. He proposed to decompose the flow of a weakly viscous fluid around an obstacle as the sum of:

- An inviscid exterior component satisfying a non penetration condition on the boundary of the obstacle.
- A boundary layer localized in the vicinity of the wall.

The role of the boundary layer is to restore the no-slip condition on the wall, which is – in general – not compatible with the inviscid equations of motion. The viscous dissipation is therefore the dominating phenomenon in the vicinity of the boundary.

The point is that this multi scale expansion is expected to provide a good approximation only for laminar flows. One can indeed observe that the boundary layer splits from the wall behind the obstacle, leading to some turbulent flow (see Fig. 3).

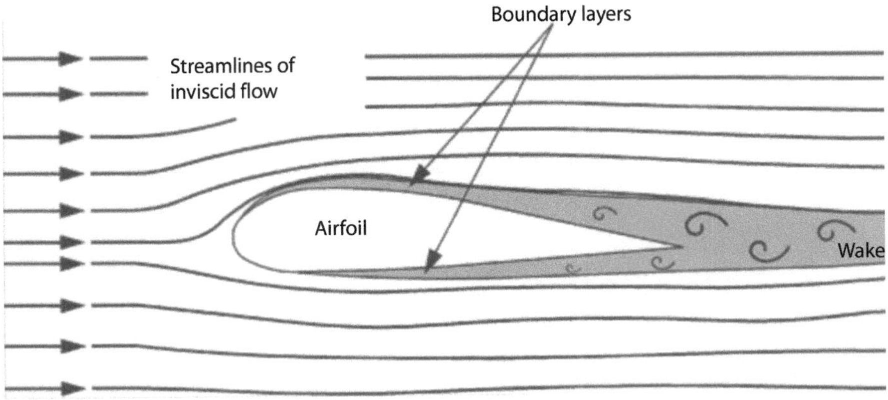

Fig. 3 Prandtl boundary layers

This phenomenon is far from being understood at the mathematical level : the only evidence of this highly nonlinear instability is the fact that the multi scale expansion does not converge [9].

3.2 Boundary layers in oceanography

A more successful mathematical analysis of boundary layers has been developed by Ekman [11] to study the transmission of surface forcing to high rotating fluids. This work aimed at explaining the observations of the explorer Nansen, who had noted that icebergs drift with an angle of 30 to 40 degrees with respect to the wind direction.

Ekman's computation, based on the balance between the Coriolis force and the viscosity, predicts an angle of 45 degrees. With depth, the current decreases and twists. This is the Ekman spiral (see Fig. 4).

The by-now classical way to establish the stability of Ekman's layers is to get a more accurate multiscale expansion with many correctors (including the Ekman suction) and to prove the convergence by energy methods. For a detailed presentation of this analysis, we refer to the books [3] and [12] where both the physical problem and the mathematical difficulties are discussed.

An alternative method, less technical insofar as it does not require the construction of so much correctors, has been developped in [21]. The method relies on weak compactness and homogenization techniques in the spirit of Allaire and N'Guetseng. The counterpart is that the asymptotics is less precisely described (no description of fast time oscillations, no rate of convergence).

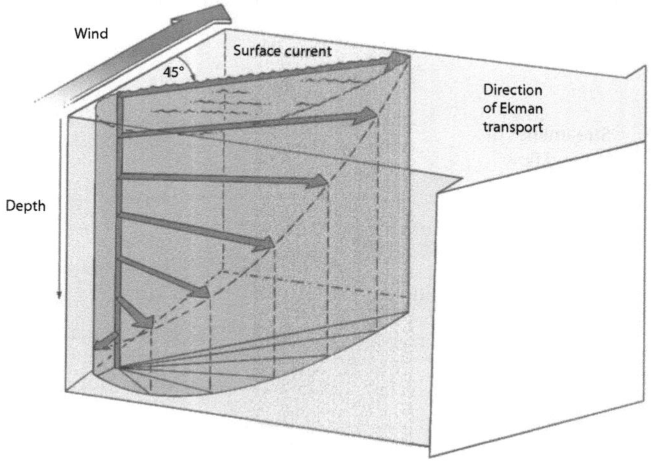

Fig. 4 The Ekman spiral. Adapted from [23]

Both methods allow to investigate more complex related problems accounting for the coupling with other physical phenomena such as the topography or the nonlinear transport. A huge literature is devoted to these problems leading to quite interesting mathematical developments. We mention for instance the influence of:

- **The roughness of the bottom.** In the case when the horizontal boundary is "rough" (periodic perturbations of a flat boundary), jump conditions must be introduced at the interface with the interior domain [13], the (linear) Ekman pumping is then replaced by some nonlinear damping term.
- **Some stochastic or periodic wind forcing.** Under appropriate non resonance assumptions [4], boundary layer correctors can be built for all Poincaré modes, which provide a suitable approximation of the fast rotating asymptotics.
  In the case of a resonant forcing, degenerate behaviours may occur (boundary layers of larger size, destabilization of the interior flow for large times...). They are characterized in [5] by the defect of hyperbolicity (in the sense of dynamical systems) of the boundary layer matrix.
- **The shallow water approximation.** The thin layer effect, when considered of the same order as the rotation (quasi-geostrophic approximation), modifies the propagation of three-dimensional Poincaré waves by creating small scales. Using tools of semiclassical analysis, one can prove that the energy propagates at speeds of order one, i.e. much slower than in traditional rotating fluid models [6]. Note that the spatial variations of the rotation vector also generate strong singularities within the boundary layer, which have repercussions on the interior part of the solution.

4 Multiscale expansion: a simple example

We now go back to the case of the two-dimensional Munk equation (1). As the velocity field u is divergence-free, one can introduce the streamfunction ψ (defined up to some additive constant) such that

$$u = \nabla^\perp \psi .$$

Taking the rotational in (1) to get rid of the pressure term, we obtain the following singular perturbation problem

$$\beta \partial_1 \psi - \nu \Delta^2 \psi = \tau \text{ in } \Omega \text{ where } \tau = \nabla^\perp \cdot \sigma,$$

It is supplemented by the boundary condition

$$\psi_{|\partial\Omega} = 0, \quad (n \cdot \nabla \psi)_{|\partial\Omega} = 0$$

denoting by n the outwards normal to $\partial\Omega$.

For the sake of simplicity, we will first exhibit the boundary layer phenomenon as well as its main qualitative features by presenting detailed computations in the 1D case. Our starting point is the singular ordinary differential equation of order 4

$$\begin{aligned} &\beta\partial_1\psi - \nu\partial_1^4\psi = \tau \text{ in }]0,1[, \\ &\psi(0) = \psi(1) = 0, \quad \psi'(0) = \psi'(1) = 0, \end{aligned} \tag{2}$$

where β and ν denote - now and in the sequel - non dimensional parameters measuring the relative influence of the restoring force due to the variations of the Coriolis parameter and of the turbulent dissipation.

4.1 The Sverdrup relation

To describe the asymptotic behaviour of ψ_ν for $\nu << 1$, we study the limit $\nu \to 0$.

Integrating by parts gives the following weighted energy estimate

$$\frac{\beta}{2}\int(\psi_\nu)^2 e^{x_1}dx_1 + \nu\int\partial_1^2(\psi_\nu)\partial_1^2(\psi_\nu e^{x_1})dx_1 = -\int\tau\psi_\nu e^{x_1}dx_1,$$

which, coupled with the Cauchy-Schwarz inequality, leads to the following uniform bound

$$\frac{\beta(1-C\nu)}{2}\int(\psi_\nu)^2 e^{x_1}dx_1 + \frac{\nu}{2}\int(\partial_1^2(\psi_\nu))^2 e^{x_1}dx_1 \le -\int\tau\psi_\nu e^{x_1}dx_1, \tag{3}$$

for some nonnegative constant C independent of ν.

In weak sense, ψ_ν therefore converges (up to extraction of a subsequence) to the solution $\bar\psi$ of the Sverdrup equation :

$$\beta\partial_1\bar\psi = \tau \text{ in }]0,1[. \tag{4}$$

Note that, at this stage, it is not clear what kind of boundary conditions can be obtained asymptotically. We only know that the Sverdrup equation, which is a hyperbolic transport equation, can admit only one boundary condition, either on the East side or on the West side but not on both (for generic source terms τ).

4.2 The boundary layer equation

The Sverdrup equation is thus not compatible with the no-slip condition. Following the strategy presented in the previous section, we therefore introduce a corrector:

$$\psi_\nu = \bar\psi + \psi_\nu^{BL}$$

which is expected to be non negligible only in the vicinity of the boundary points.

More precisely, we require that:

- the boundary layer restores boundary conditions

$$\bar\psi(0) = -\psi_\nu^{BL}(0), \quad \bar\psi(1) = -\psi_\nu^{BL}(1),$$

$$\frac{d}{dx_1}\bar\psi(0) = -\frac{d}{dx_1}\psi_\nu^{BL}(0), \quad \frac{d}{dx_1}\bar\psi(1) = -\frac{d}{dx_1}\psi_\nu^{BL}(1);$$

- it is dominated by viscous effects

$$\beta\partial_1\psi_\nu^{BL} - \nu\partial_1^4\psi_\nu^{BL} = 0.$$

Let us indeed recall that the Sverdrup equation already accounts for the wind forcing.

4.3 East/West disymmetry

The thickness of the layer is given by the scaling. In this very simple setting, we have

$$\psi_\nu^{BL}(x_1) = \psi^E\left(\frac{1-x_1}{(\nu/\beta)^{1/3}}\right) + \psi^W\left(\frac{x_1}{(\nu/\beta)^{1/3}}\right).$$

- In the East, decaying solutions are of the form

$$\psi^E(z) = \lambda\exp(-z).$$

- The space of West solutions is of dimension 2

$$\psi^W(z) = \lambda_1\exp(jz) + \lambda_2\exp(j^2z)$$

denoting by j, j^2 the non real cubic roots of unity.

The boundary condition for the Sverdrup equation is therefore prescribed on the East side. The constants $\bar\psi(1)$, λ, λ_1 and λ_2 are then determined by solving a linear system of algebraic relations

$$\lambda + \bar\psi(1) = 0, \quad \lambda\left(\frac{\beta}{\nu}\right)^{1/3} + \frac{\tau(1)}{\beta} = 0,$$

$$\lambda_1 + \lambda_2 + \bar\psi(0) = 0, \quad -\lambda_1 j\left(\frac{\beta}{\nu}\right)^{1/3} - \lambda_2 j^2\left(\frac{\beta}{\nu}\right)^{1/3} + \frac{\tau(0)}{\beta} = 0,$$

$$\bar\psi(0) = \bar\psi(1) - \int_0^1 \frac{\tau(s)}{\beta}ds$$

For smooth τ, we check that the approximation is consistent, or in other words that $\bar\psi + \psi_\nu^{BL}$ satisfies the Munk equation with small error terms, as well as the exact boundary conditions. A refinement of the energy estimate (3) then shows that

$$\|\psi_\nu - \bar\psi - \psi_\nu^{BL}\|_{L^2} << \|\bar\psi + \psi_\nu^{BL}\|_{L^2}$$

meaning that $\bar\psi + \psi_\nu^{BL}$ is indeed the leading order approximation of ψ_ν.

5 Influence of the geometry: some remarkable features

Our goal in [7] was to further understand the influence of the geometry of the 2D domain Ω on this kind of multiscale expansion. More specifically, we were interested in describing the behaviour in the vicinity of North and South coasts where the previous analysis fails, which was let as a remaining open problem in the seminal paper [8] by Desjardins and Grenier on Munk boundary layers. They indeed had to assume that the forcing vanishes in the vicinity of the poles.

5.1 Northern/Southern degeneracy

We first note that in the vicinity of North and South boundaries, the transport term is not singular :

$$\beta \partial_1 \psi - \nu \partial_2^4 \psi = 0. \tag{5}$$

We therefore expect the *size of the boundary layer* to be different

$$\psi_\nu^{BL}(x_1, x_2) = \psi^N\left(x_1, \frac{1-x_2}{(\nu/\beta)^{1/4}}\right) + \psi^S\left(x_1, \frac{x_2}{(\nu/\beta)^{1/4}}\right).$$

Furthermore, the boundary layer equation becomes a *diffusion-like equation*, with the arc-length s playing the role of the time variable. Note that such degenerate parabolic boundary layers have been exhibited in [10] for instance.

What can actually be proved is that the non local equation

$$\begin{aligned}
&\partial_s g + \partial_z^4 g = 0, \quad s \in (0,T),\ z > 0,\\
&g_{|s=0} = g_{in}\\
&g_{|Z=0} = g_0, \quad \partial_z g_{|Z=0} = g_1,
\end{aligned}$$

with $g_{in} \in L^2(0,\infty)$, $g_0, g_1 \in W^{1,\infty}(0,T)$, $b \in L^\infty((0,T), W^{1,\infty}(\mathbf{R}_+))$ is well-posed, and that its solutions satisfy some weighted energy estimates implying in particular that they remain localized in the vicinity of the boundary.

In the present context, this means that on South and North boundary layers, equation (5) is a backward equation (in x_1). This is consistent with the definition of the interior term $\bar{\psi}$: in all cases, the boundary condition is prescribed on the East end of the interval.

5.2 Discontinuity zones

For complex domains, i.e. for domains where the closure of the East boundary $\bar{\Gamma}_E$ is not connected, the solution of the Sverdrup equation $\bar{\psi}$ has discontinuities across the horizontal lines Σ_{ij} (see Fig. 5).

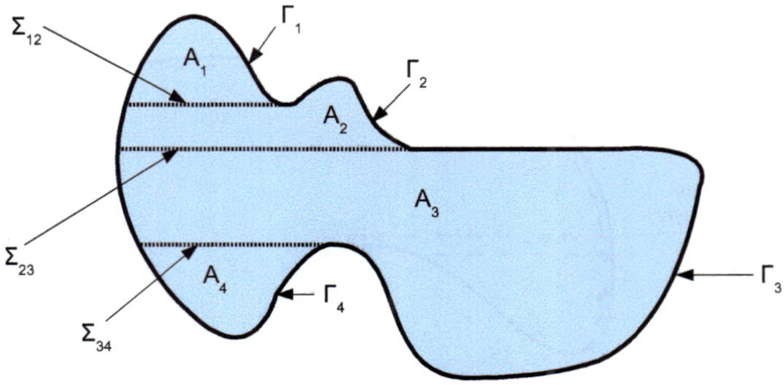

Fig. 5 Discontinuities of the Sverdrup flow

For the exposition, we will assume here that there is only one such discontinuity line Σ, say at longitude y_2.

Denote by y_1 the abscissa of the singular point of the boundary. For $x_1 < y_1$,

$$[\bar{\psi}]_{|\Sigma}(x_1,y_2) = \bar{\psi}(x_1,y_2^+) - \bar{\psi}(x_1,y_2^-) = -\int_{x_E^-(y_2)}^{x_E^+(y_2)} \tau(x_1,y_2)\,dx_1.$$

Notice in particular that the jump is constant along Σ.

Up to some suitable truncation, we can assume that $\tau(x_E^-(y_2),y_2)$ vanishes in a neighbourhood on the left of y_2, so that we get in a similar way

$$[\partial_2\bar{\psi}]_{|\Sigma}(x_1,y_2) = -(x_E^+)'(y_2)\tau(x_E^+(y_2),y_2) - \int_{x_E^-(y_2)}^{x_E^+(y_2)} \partial_2\tau(x_1,y_2)\,dx_1.$$

The next step is to construct some lifting term to counterbalance the jump of $\bar{\psi}$, and of its normal derivative $\partial_2\bar{\psi}$, across Σ:

$$[\psi_l]_{l\Sigma} = -[\bar{\psi}]_\Sigma, \quad [\partial_2\psi_l]_{l\Sigma} = -[\partial_2\bar{\psi}]_\Sigma.$$

When we lift these boundary conditions, we introduce a source term in the equation. This source term is then handled by a boundary layer type term ψ^Σ, which has no discontinuity across Σ.

These discontinuities thus gives rise to a *boundary layer singularity*, which is apparented to the North and South boundary terms, the size of which is therefore $\nu^{1/4}$. If we could have nice formulations for the North and South boundary layer operators, we would expect to get $(\psi_l + \psi^\Sigma)$ by a rather simple process, obtained by application of the previous results. Nevertheless, as:

- the boundary layer equation is highly nonlocal (propagation with respect to s);
- the fictitious boundary is not perfectly horizontal,

we are not able to proceed in such a way.

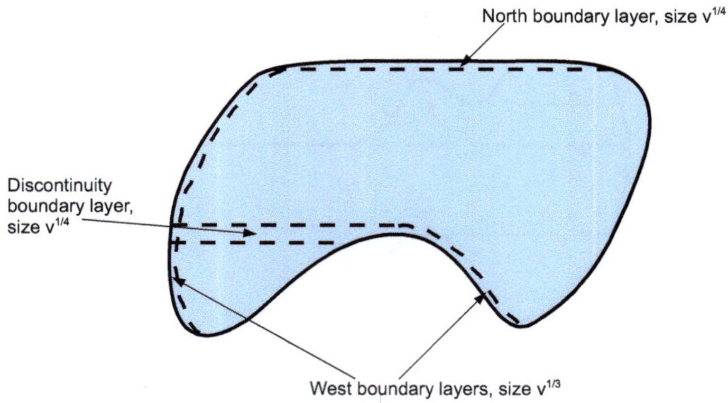

Fig. 6 Boundary layers for the Munk equation

Remark 5.1 *If the domain Ω has islands, we proceed in a very similar way adding to $\bar{\psi}$ some prescribed quantity, which corresponds to fixing the circulation around these islands.*

5.3 Complex transitions

The approximate solutions to the Munk equation (1) are obtained by gathering together the different elementary pieces described previously, namely:

- the interior term which is essentially the solution to the transport equation (4), regularized in the vicinity of "East corners" and of the interface Σ;
- East and West boundary layer terms, which lift locally the boundary conditions but become singular in the vicinity of the corners;
- North and South boundary layer terms, which lift the boundary conditions on the horizontal parts of the boundary but in a non local way.

The major difficulty is to understand the interplay between those different elementary pieces (see Fig. 6). Of course, the equation (1) being linear, the errors induced by all these terms are simply added, so that the control on the remainders in the approximate equation will be rather simple to obtain. The point to be stressed is that we need that each elementary term to be smooth enough (namely H^2), with suitable controls on the corresponding derivatives.

We will not enter the details of this technical part. Let us just mention some important points to be noted insofar as they differ from usual boundary layer techniques.

The first remark is that one has no matching of boundary layers on the corners (which would not be possible anyway since they have different sizes), but rather a *superposition*.

The second point is related to the fact that North and South boundary layers are defined by a non local equation. One has therefore to establish some *extinction prop-*

erty, that is check that they do not carry any more energy beyond some point, so that they can be truncated and considered as local contributions.

Finally, we would like to emphasize that the construction must be performed in a precise order, starting with East boundaries, then defining North/South (and surface) boundary layers, and finally lifting the West boundary conditions. This *dissymmetry between East and West boundaries* is similar to what happens at the macroscopic level for the interior term.

Starting from the scaled Munk equation (1), we therefore end up with a very precise description of the asymptotics $\nu \to 0$, which requires to build a variety of correctors. A natural question is to know whether these techniques are robust when considering more complex models (involving typically systems of equations) for which the same semi-explicit computations cannot be done. In particular, understanding the influence of temperature and stratification is a major challenge.

References

1. Bresch, D., Colin, T.: Some remarks on the derivation of the Sverdrup relation. J. Math. Fluid Mech. **4**(2), 95–108 (2002)
2. Bresch, D., Guillén-Gonzalez, F., Rodríguez-Bellido, M.A.: A corrector for the Sverdrup solution for a domain with islands. Appl. Anal. **83**(3), 217–230 (2004)
3. Chemin, J.-Y., Desjardins, B., Gallagher, I., Grenier, E.: Basics of Mathematical Geophysics. Oxford Lecture Series in Mathematics and its Applications **32**. Oxford University Press, New York (2006)
4. Dalibard, A.-L. : Asymptotic behavior of a rapidly rotating fluid with random stationary surface stress. SIAM J. Math. Anal. **41**, 511–563 (2009)
5. Dalibard, A.-L., Saint-Raymond, L.: Mathematical study of rotating fluids with resonant surface stress. J. Differential Equations **246**(6), 2304–2354 (2009)
6. Dalibard, A.-L., Saint-Raymond, L.: Mathematical study of the β-plane model for rotating fluids in a thin layer. J. Math. Pures Appl. **94**(2), 131–169 (2010)
7. Dalibard, A.-L., Saint-Raymond, L.: Mathematical study of degenerate boundary layers. Preprint (2012). arxiv 1203.5663
8. Desjardins, B., Grenier, E.: On the Homogeneous Model of Wind-Driven Ocean Circulation, SIAM Journal on Applied Mathematics **60**, 43–60 (1999)
9. Gérard-Varet, D., Dormy, E.: On the ill-posedness of the Prandtl equation. J. Amer. Math. Soc. **23**, 591–609 (2010)
10. Eckhaus, W., de Jager, E.M.: Asymptotic solutions of singular perturbation problems for linear differential equations of elliptic type. Arch. Rational Mech. Anal. **23**, 26–86 (1966)
11. Ekman, V.W.: On the influence of the earth's rotation on ocean currents. Ark. Mat. Astron. Fys. **2**, 1–52 (1905)
12. Gallagher, I., Saint-Raymond, L.: On the influence of the Earth's rotation on geophysical flows. In: Friedlander, S., Serre, D. (eds.) Handbook of Mathematical Fluid Dynamics. Elsevier, The Netherlands (2007)
13. Gérard-Varet, D.: Highly rotating fluids in rough domains. J. Math. Pures Appl. **82**, 1453–1498 (2003)
14. Gill, A. E.: Atmosphere-Ocean Dynamics. International Geophysics Series, Vol. 30. Academic Press, London (1982)
15. Grasman, J.: On the birth of boundary layers. Mathematical Centre Tracts 36. Mathematisch Centrum, Amsterdam (1971)

16. Grenier, E.: Oscillatory perturbations of the Navier–Stokes equations. Journal de Mathématiques Pures et Appliquées **76**, 477–498 (1997)
17. Mariano, Global Surface Velocity Analysis (MGSVA) 1.0, US Coast Guard's (1995)
18. Pedlosky, J.: Geophysical fluid dynamics. Springer-Verlag, New York (1979)
19. Pedlosky, J.: Ocean Circulation Theory. Springer-Verlag, Berlin Heidelberg (1996)
20. Prandtl, L.: Über Flüssigkeitsbewegung bei sehr kleiner Ribung. Verh. III. Intern. Math. Kong., pp. 484–491. Heidelberg (1905)
21. Saint-Raymond, L.: Weak compactness methods for singular penalization problems with boundary layers. SIAM J. Math. Anal. **41**(1), 153–177 (2009)
22. Schochet, S.: Fast singular limits of hyperbolic PDEs. Journal of Differential Equations **114**, 476–512 (1994)
23. Thurman, H.Y.: Essentials of oceanography, 5th ed. Prentice Hall, Inc. (1996)

Noise-induced Periodicity: Some Stochastic Models for Complex Biological Systems

Paolo Dai Pra, Giambattista Giacomin and Daniele Regoli

Abstract After a review of some examples of life science stochastic models, we propose a stylized model with characteristics inspired by the examples above, reproducing noise-induced pulsations as a collective macroscopic phenomenon.

1 Introduction

Stochastic models with many interacting degrees of freedom, which have been originally mainly inspired by Statistical Physics, are nowadays widely spread in different disciplines. A largely incomplete list of applications is the following:

- Social Sciences: multi-agent models, systemic risk, contagious default, opinion dynamics.
- Engineering: distributed control, communication networks, quantum networks.
- Biology: species competitions, networks of neurons, multicellular structures.

Although stochastic models inspired by Statistical Physics are useful in all these contexts, complex systems in social sciences and biology often exhibit peculiar features and patterns that are not accounted for by traditional models in Statistical Physics. This is directly related to the non reversible nature – i.e. absence of detailed balance – of most life science stochastic models. Just to cite an example close to the

P. Dai Pra (✉)
Dipartimento di Matematica, Università degli Studi di Padova, via Trieste 63, I-35121 Padova, Italy
e-mail: daipra@math.unipd.it

G. Giacomin
Université Paris Diderot, U.F.R. de Mathématiques, Bât. Sophie Germain, 5 rue Thomas Mann, 75205 PARIS Cedex 13, France
e-mail: giambattista.giacomin@univ-paris-diderot.fr

D. Regoli
Dipartimento di Matematica, Università degli Studi di Padova, via Trieste 63, I-35121 Padova, Italy
e-mail: regoli@math.unipd.it

A. Celletti, U. Locatelli, T. Ruggeri and E. Strickland (eds.): *Mathematical Models and Methods for Planet Earth.* Springer INdAM Series 6, DOI 10.1007/978-3-319-02657-2_3,

content of this note, the prototypical model for synchronization phenomena, the Kuramoto model [1], is reversible and coincides with a classical statistical mechanics model only in a very particular case [2], that is when the natural frequencies of all units coincide, and this particular case misses a fundamental modeling character and the connected relevant phenomenology. Our purpose in this paper is to discuss a special but important phenomenon, that we will refer to as *noise-induced periodicity*.

It is well known that complex biological systems often exhibit organized pulsating behavior; concerning humans, beside obvious examples such as heartbeat and respiration, periodic behavior in the electrical activity of brain neurons has been observed, and its relation with pathologies such as epilepsy has been investigated [3, 12, 13]. These observations have stimulated the formulation of new models, in the attempt of identifying the basic causes giving origin to large-scale organized behavior, and of determining those *universal* features that are robust with respect to the details of the model.

Without making any attempt of being exhaustive, we begin by reviewing four basic models which are of interest in this context; we briefly describe their properties and fields of applications. All these models share the feature of being defined at a microscopic/mesoscopic level, i.e. at the level of single cells or neurons.

1.1 A simple genetic circuit: the repressilator

The repressilator is a model for gene expression. For a single cell it consists of a network of three genes, whose expression depends on the concentration of their related proteins, in particular the increase in the concentration of the proteins inhibits the production of each other in a cyclic manner. Denoting by x, y, z the concentrations of the three proteins in a given cell, chemical reactions produce dynamics for these concentrations described by a nonlinear differential system (see e.g. [7, 8]):

$$\begin{aligned} \dot{x} &= -x + \frac{\alpha}{1+z^n} \\ \dot{y} &= -y + \frac{\alpha}{1+x^n} \\ \dot{z} &= -z + \frac{\alpha}{1+y^n} \end{aligned} \tag{1}$$

where $\alpha > 0$ and $n \geq 1$ are the parameters of the model. This evolution has a unique limit cycle: in the steady state, proteins' concentrations pulsate periodically [7].

In a multicellular system, denote by x_i, y_i, z_i the concentrations in the ith cell, $i = 1, 2, \ldots, N$. The dynamics may be affected by special molecules (known as autoinducer (AI) molecules) which diffuse through cellular membranes. In the intracellular space, AI molecules react with the gene-related proteins; in the extracellular space they are subject to degradation, and diffuse fast, so that their extracellular concentration can be regarded as constant in space. If we denote by s_i the AI concentration in the ith cell, and by S the extracellular concentration, the joint dynamics is given by

$$\begin{aligned}
\dot{x}_i &= -x_i + \frac{\alpha}{1+z_i^n} \\
\dot{y}_i &= -y_i + \frac{\alpha}{1+x_i^n} \\
\dot{z}_i &= -z_i + \frac{\alpha}{1+y_i^n} + \frac{\beta s_i}{1+s_i} \\
\dot{s}_i &= -h_0 s_i + h_1 x_i + \gamma(S - s_i) \\
\dot{S} &= -kS - \delta\left(S - \frac{1}{N}\sum_{i=1}^{N} s_i\right),
\end{aligned} \tag{2}$$

where all parameters are strictly positive. Different cells interact indirectly through the AI molecules; due to fast extracellular diffusion, the interaction is of *mean-field* type: spatial position of cells does not matter. This model is deterministic, but it has been observed that noise should be added to account for interactions with the environment, thus making the model more realistic. Numerical simulations, both in the deterministic model and in its stochastic modification, suggest that for β and δ large enough, pulsations in cellular protein concentrations synchronize, producing collective periodic behavior. This is in agreement with various experiments on real cellular cultures (see also [11]).

Moreover a rigorous result establishing large scale periodic behavior for *interacting noisy Brusselators* has been established in [17] (the Brussellator is a system of coupled ODEs with a stable limit cycle, like the the repressillator).

1.2 Interacting rotators: the random Kuramoto model

We have seen that in the repressilator model an isolated cell behaves periodically in time, in the steady regime. A substantial effort has been put into understanding to which extent the essential features of systems of this type can be captured by systems of *interacting rotators*, a procedure that goes under the name of *phase reduction*. The most popular model in this context is due to Kuramoto [10]. Denote by $\theta_i \in [0, 2\pi)$, $i = 1, 2, \ldots, N$, the phase of the ith rotator. In the following version of the model, the interaction among rotators is of mean-field type, and the dynamics of each rotator is affected by a Brownian noise:

$$\mathrm{d}\theta_i(t) = \omega\,\mathrm{d}t + \frac{\beta}{N}\sum_{j=1}^{N}\sin(\theta_j(t) - \theta_i(t))\,\mathrm{d}t + \sigma\,\mathrm{d}W_i(t), \tag{3}$$

where ω is the characteristic frequency of the rotators, $\beta > 0$ controls the strength of the interaction, which has the tendency of synchronizing the oscillators, $\sigma > 0$ denotes the noise intensity, and $(W_i)_{i=1}^N$ are independent standard Brownian motions.

To fix notations, Brownian motions are defined on a probability space $(\Omega, \mathscr{F}, \mathbf{P})$. For a real o complex valued random variable X, we denote by $\mathbf{E}(X) := \int X d\mathbf{P}$ its expectation with respect to $\mathbf{P}$. Similar notations will be used for other stochastic models in this paper.

It is known (see [1]) that this model may exhibit a *synchronization phase transition*. In order to describe this phase transition, one notes that, in the limit as $N \longrightarrow +\infty$, the macroscopic behavior of the ensemble of rotators is described by the mean-field equation for the evolution of a "typical" rotator

$$d\theta(t) = \omega\, dt + \beta \int \sin(\theta' - \theta(t))\mu_t(d\theta') + \sigma\, dW(t), \tag{4}$$

where μ_t is the distribution of $\theta(t)$. The term $\beta \int \sin(\theta' - \theta(t))\mu_t(d\theta')$ can be interpreted as the *effective field* felt by one rotator due to the interaction with the rest of the ensemble. An effective way to study the distribution of solutions of (4), is to introduce the *order parameters* r_t, ψ_t by

$$r_t e^{i\psi_t} := \mathbf{E}\left[\exp\{i(\theta(t) - \omega t)\}\right].$$

Fixing the parameters ω and σ, for β below a critical value $\beta_c > 0$, it can be shown that all solutions of (4) are such that

$$\lim_{t \longrightarrow +\infty} r_t = 0.$$

In other words, in the steady state regime, phases have a incoherent distribution, producing a null macroscopic effect: the rotators are *de-synchronized*. For $\beta > \beta_c$ de-synchronized solutions of (4) still exist, but they are unstable. Excluding these unstable solutions, for all other solutions the limit

$$re^{i\psi} := \lim_{t \longrightarrow +\infty} r_t e^{i\psi_t}$$

exists, and $r \neq 0$: rotators synchronize, producing pulsations with frequency ω at macroscopic level.

Collective periodic behavior is thus the result of synchronization and the fact that rotators have the same characteristic frequency. As observed in [16, 19], collective periodicity may emerge in a more subtle way in a modification of the above model, tackled mathematically in [9]. Assume $\omega = 0$, so no "natural" rotation is present. Equation (4), for $\beta > \beta_c$, has a one dimensional family of stationary, synchronized solutions, which differ by a phase shift. Consider a small perturbation of (3), obtained by adding a *self-interaction* term of the form $\varepsilon U(\theta_i(t))$, where ε is a small parameter and U a suitable regular function (rotators are called *active* in this model). At macroscopic level, this amounts to replace (4) by

$$d\theta(t) = \varepsilon U(\theta(t))\, dt + \beta \int \sin(\theta' - \theta(t))\mu_t(d\theta') + \sigma\, dW(t). \tag{5}$$

For ε sufficiently small, it can be shown that (5) (more precisely the flow of the distributions μ_t) has a one dimensional invariant manifold, whose elements, for $\varepsilon > 0$, are not fixed points. For suitable choices of $U(\cdot)$ this gives rise to stable periodic trajectories. Note that this periodic behavior emerges even though there is no intrinsic periodicity of a single rotator, and it is therefore a purely collective phenomenon.

1.3 A model for neurons interaction

In a minimal description of neuron's activity, the state of a neuron is represented by a real number V, its *firing rate*. In absence of external signals this rate decays to zero exponentially. In a large ensemble of N neurons, each neuron receives signals from others; this interaction can be modeled as an effective drift on the firing rate V_i of the ith neuron, proportional to a *sigmoidal* function $S(V_j)$ of the firing rate of every other neuron j. Thus, in the *mean-field* approximation, and assuming the presence of Brownian noises affecting the dynamics, one obtains the system of stochastic differential equations

$$\mathrm{d}V_i(t) = -\alpha V_i(t)\,\mathrm{d}t + \frac{J}{N}\sum_{j=1}^{N} S(V_j(t))\,\mathrm{d}t + \sigma\,\mathrm{d}W_i(t), \tag{6}$$

where $(W_i)_{i=1}^N$ are independent standard Brownian motions, J controls the interaction strength, and σ is the level of noise.

In the $N \to +\infty$ limit, the macroscopic ensemble behavior is described by the dynamics of a "typical" neuron:

$$\mathrm{d}V(t) = -\alpha V(t)\,\mathrm{d}t + J\mathbf{E}[S(V(t))]\,\mathrm{d}t + \sigma\,\mathrm{d}W(t). \tag{7}$$

Note that, by linearity of this equation in V, if $V(0)$ has a Gaussian distribution then $V(t)$ is Gaussian for all times. Mean and variance, as functions of t, solve decoupled equations; the variance solves a linear equation, which does not depend on the term $\mathbf{E}[S(V(t))]$, so the only effective variable is the mean, which solves a nonlinear ordinary diffential equation.

In [20] an extension of this model is considered. Neurons are not treated as all equal, but each belongs to one of d different classes, each one comprised by a large number N_k of neurons, $k = 1,\dots,d$. The interaction strength between two neurons depends on the class they belong to. Consider, for simplicity, the case $d = 2$, and denote by $(U_i)_{i=1}^{N_1}$, $(V_j)_{j=1}^{N_2}$ the firing rates of the neurons of the two classes. In this generalization, equation (6) is replaced by the system

$$\begin{aligned}
\mathrm{d}U_i(t) &= -\alpha U_i(t)\,\mathrm{d}t + \frac{J_{11}}{N_1}\sum_{h=1}^{N_1} S(U_h(t))\,\mathrm{d}t + \frac{J_{12}}{N_2}\sum_{k=1}^{N_2} S(V_k(t))\,\mathrm{d}t + \sigma\,\mathrm{d}W_i^{(1)}(t) \\
\mathrm{d}V_j(t) &= -\alpha V_j(t)\,\mathrm{d}t + \frac{J_{21}}{N_1}\sum_{h=1}^{N_1} S(U_h(t))\,\mathrm{d}t + \frac{J_{22}}{N_2}\sum_{k=1}^{N_2} S(V_k(t))\,\mathrm{d}t + \sigma\,\mathrm{d}W_j^{(2)}(t),
\end{aligned} \tag{8}$$

where $(\mathrm{d}W_i^{(1)})_{i=1}^{N_1}$, $(\mathrm{d}W_j^{(2)})_{j=1}^{N_2}$ are independent Brownian motions, and the macroscopic equation (7) is replaced by

$$\begin{aligned} \mathrm{d}U(t) &= -\alpha U(t)\,\mathrm{d}t + J_{11}\mathbf{E}[S(U(t))]\,\mathrm{d}t + J_{12}\mathbf{E}[S(V(t))]\,\mathrm{d}t + \sigma\,\mathrm{d}W^{(1)}(t) \\ \mathrm{d}V(t) &= -\alpha V(t)\,\mathrm{d}t + J_{21}S(U(t))\,\mathrm{d}t + J_{22}\mathbf{E}[S(V_k(t))]\,\mathrm{d}t + \sigma\,\mathrm{d}W^{(2)}(t). \end{aligned} \tag{9}$$

As in the $d = 1$ case, this system is *Gaussian*, and it can be reduced to a two-dimensional system of ordinary differential equations. It is shown in [20] that periodic solutions may emerge in this latter equation, depending on the parameters of the model. In particular, it is shown that for certain fixed values of the coupling constants J_{ab} and choice of the function $S(\cdot)$, periodic solutions surprisingly appear *by increasing the noise level* σ. This phenomenon is referred to as *noise induced periodicity*.

1.4 Active Brownian particles

The following model (see e.g. [18]) has been proposed to describe the motion of family of cells that are drifted towards higher concentration of a given chemical. Cells are *active* in that they themselves release the chemical. If $h(x,t)$ denotes the concentration of the chemical at $x \in \mathbf{R}^3$ and at time t, the position X_i of the ith cell evolves according to the stochastic differential equation

$$\mathrm{d}X_i(t) = \partial_x h(X_i(t),t)\,\mathrm{d}t + \sigma\,\mathrm{d}W_i(t), \tag{10}$$

where $(W_i)_{i=1}^N$ are independent 3-dimensional Brownian motions, and $\sigma > 0$ is a scalar. The concentration $h(x,t)$ has its own evolution:

$$\partial_t h(x,t) = -\alpha h(x,t) + D\Delta h(x,t) + \beta \sum_{j=1}^{N} g(X_j(t),x), \tag{11}$$

which is subject to dissipation ($\alpha > 0$), diffusion ($D > 0$), while the third summand describes the coupling with the cells that, releasing the chemical, modify its concentration.

This model, that has been studied mostly on the basis of numerical simulations, exhibits a very rich spatio-temporal behavior, which includes spikes formation and pulsations. For more details on this and related models we refer to [14]. For our purposes, the key property of this model is the fact that particles contribute to the potential driving their own motion.

2 A Curie-Weiss model with dissipation and noise

After having reviewed some of the basic models in cellular and neural dynamics, we propose here a model whose state variables are $\{-1,1\}$ – valued spins. The main purpose is to reproduce, although at a stylized level, some of the features of the cellular dynamics in Section 1.4, while keeping in part the analytic tractability of the neural model in Section 1.3. Moreover, the mechanism producing collective oscillations is reminiscent of those of the active rotators in Section 1.2.

We consider a system of N spins $s = (s_1, s_2, \dots, s_N) \in \{-1,1\}^N$. Each spin is subject to a possibly time dependent local field $\lambda_i \in \mathbf{R}$. The continuous-time dynamics, which describes the tendency of the spin to align to their local field, is the following: each transition $s_i \to -s_i$ occurs at rate $k[1-\tanh(\lambda_i s_i)]$, $k > 0$. In more probabilistic terms, this dynamics is realized by introducing driving Poisson processes of intensity $2k$, i.e. a family $\{\tau_j^{(i)} : j \geq 0, i = 1, \dots, N\}$ of random times, with $\tau_0^{(i)} = 0$ and such that $\{\tau_{j+1}^{(i)} - \tau_j^{(i)} : j \geq 0, i = 1, \dots, N\}$ are independent, positive random variables with $\mathbf{P}(\tau_{j+1} - \tau_j > h) = \exp(-2kh)$. At each time $t = \tau_j^{(i)}$, a coin is thrown, with probability of head equal to $(1-\tanh(\lambda_i(t)s_i(t)))/2$. If a head appears, then the transition $s_i \to -s_i$ occurs, otherwise the spin is left unchanged. The parameter k is thus interpreted as the intensity of the Poissonian noise driving this dynamics.

As for the model in Section 1.4 spins are *active*, in the sense that they modify their local field: the $\lambda_i(t)$'s are themselves stochastic processes, evolving according to the stochastic differential equations

$$\mathrm{d}\lambda_i(t) = -\alpha\lambda_i(t)\,\mathrm{d}t + \sigma\sqrt{k}\,\mathrm{d}B_i(t) + \beta\,\mathrm{d}m_N(s(t)), \tag{12}$$

where $\alpha, \beta, \sigma \geq 0$,

$$m_N(s) := \frac{1}{N}\sum_{k=1}^{N} s_k, \tag{13}$$

and with $B_1, B_2, \dots, B_N$ independent Brownian motions. The observable m_N is called *magnetization*. Thus, when the ith spin jumps, all local fields increase by $-2\beta s_i/N$: this introduce a tendency of the spins to align, similar to what happens in ferromagnets. Unlike in usual stochastic dynamics for ferromagnets, the local fields are not deterministic functions of the spin vector, but may be subject to dissipation ($\alpha > 0$) and Brownian noise ($\sigma > 0$). The choice of the factor $\sqrt{k}$ in the Brownian term in (12) is such that two sources of noise scale consistently in time: the time change $t \to kt$ would correspond to the change of parameters $k \to 1$, $\alpha \to \alpha/k$, while β, σ are left unchanged.

Notice that, if we cancel out dissipation and Brownian noise, i.e. we put $\alpha = \sigma = 0$, we are left precisely with a model for ferromagnets, namely a (modification of) Curie-Weiss model, with an external magnetic field depending on the initial condition of the local fields.

Similarly to what we have seen in (4), (7) and (9), in the limit as $N \to +\infty$ the dynamics of the ensemble is described by the mean-field equation

$$\begin{cases} \Sigma(t) \to -\Sigma(t) \quad \text{with intensity} \quad k[1-\tanh(\Sigma(t)\Lambda(t))]\,, \\ \mathrm{d}\Lambda(t) = -\alpha\Lambda_t\,\mathrm{d}t + 2\beta k\mathbf{E}\left[\tanh(\Lambda(t)) - \Sigma(t)\right]\mathrm{d}t + \sigma\sqrt{k}\,\mathrm{d}B_t. \end{cases} \tag{14}$$

Alternatively, denoting formally by $p_t(s,\lambda)\,\mathrm{d}\lambda$ the joint distribution of (Σ_t,Λ_t), p_t can be identified as the weak solution of the nonlinear equation

$$\partial_t p_t(s,\lambda) = k(1+s\tanh(\lambda))p_t(-s,\lambda) - k(1-s\tanh(\lambda))p_t(s,\lambda) + \frac{k\sigma^2}{2}\partial_\lambda^2 p_t(s,\lambda) - 2\beta k g(t)\partial_\lambda p_t(s,\lambda) + \alpha\partial_\lambda(\lambda p_t(s,\lambda)) \,, \quad (15)$$

where

$$g(t) := \sum_s \int_{-\infty}^{+\infty} p_t(s,\lambda)(\tanh(\lambda)-s)\,\mathrm{d}\lambda \,.$$

For $\sigma > 0$, p_t is indeed a classical solution of (15).

Our main purpose is to determine the long-time behavior of the solutions of either (14) or (15), in particular detect stable fixed points and periodic solutions.

2.1 The case of no Brownian noise

We begin by considering the special case in which $\sigma = 0$ and $\Lambda(0)$ is deterministic: in other words, there is no Brownian noise, and the ensemble of the initial fields is concentrated near a given value. In this case (see [5]) the full phase diagram of the stationary solutions of (15) can be determined. Indeed, in analogy with what happens for the model in Section 1.3, (15) can be reduced to the following system of two ordinary differential equations for $\lambda(t) := \mathbf{E}(\Lambda(t))$ and $m(t) := \mathbf{E}(\Sigma(t))$:

$$\begin{cases} \dot\lambda(t) = 2\beta k(\tanh(\lambda(t)) - m(t)) - \alpha\lambda(t) \,, \\ \dot m(t) = 2k(\tanh(\lambda(t)) - m(t)). \end{cases} \quad (16)$$

This nonlinear system is of the *Liénard* type (see, for example, [4, 15]). Classical results on these systems yield the following

Theorem 1 *Assume* $\alpha > 0$.

(i) *For* $\beta \le \frac{\alpha}{2k}+1$ *the origin is a global attractor for* (16).

(ii) *For* $\beta > \frac{\alpha}{2k}+1$ *the system* (16) *has a unique periodic orbit, which attracts all trajectories except the fixed point, which becomes unstable.*

The fixed point $(m,\lambda) = (0,0)$ corresponds to a *asynchronized* ensemble of spins: the spins are evenly distributed to produce a null magnetization. This state attracts all initial conditions for $\beta \le \frac{\alpha}{2k}+1$, in particular whenever $\beta \le 1$. For $\beta > 1$, if k is sufficiently large compared to α, spins synchronize to produce a nonzero magnetization. However, unlike in Ising-type ferromagnets, the magnetization is not constant in time, and exhibits a periodic behavior.

The emergence of pulsations has some common features with the active rotators in Section 1.2, at least for small values of the dissipation parameter α. As for the $\varepsilon = 0$ case in that model, for $\alpha = 0$ in the system (16), all points in the one dimensional manifold $m = \tanh(\lambda)$ are fixed. Unlike in the case of rotators, however, for

$\beta > \frac{\alpha}{2k} + 1$ this manifold has an unstable component near the origin, and two stable branches. A small dissipation introduces a slow motion on the stable part of the manifold; as the unstable part is met, the motion is driven to the other stable branch. This scheme is then repeated, producing periodic motion. We remark that analogous properties are found in the classical Van der Pol oscillator, whose equation is indeed quite similar to (16); in particular, as β crosses the critical value $\frac{\alpha}{2k} + 1$, a Hopf bifurcation occurs.

2.2 *The effects of the Brownian noise*

As we allow $\sigma > 0$, the full infinite dimensional dynamics (15) has to be studied, and much less is understood on the steady state solutions. An asynchronous stationary solution $\overline{p}$ can be shown to exist, i.e. a stationary solution $\overline{p}(s,\lambda)$ of (15) for which $\int \overline{p}(s,\lambda)\,d\lambda = \frac{1}{2}$ for $s = \pm 1$ and whose λ-marginal is given by

$$\sum_s \overline{p}(s,\lambda) = \sqrt{\frac{\alpha}{\pi\sigma^2}} \exp\left(-\frac{\alpha\lambda^2}{\sigma^2}\right).$$

Stability of this solution, and the existence of periodic orbits, are problems that are not yet fully understood [6].

Via a perturbation argument, we can show that the existence of periodic solutions is preserved for small values of σ.

Theorem 2 *Assume* $\beta > \frac{\alpha}{2k} + 1$. *Then there exists* $\underline{\sigma} = \underline{\sigma}(\alpha,\beta,k) > 0$ *such that for all* $\sigma < \underline{\sigma}$ *Equation* (15) *has a non-constant periodic solution, and the asynchronous solution* $\overline{p}$ *is unstable.*

Even in this perturbative regime, however, we cannot control uniqueness and stability of periodic solutions.

A relevant fact, suggested by heuristics and rigorously proved to some extent, is that the Brownian noise plays against synchronization; this differs from the effects of the Possonian noise in the $\sigma = 0$ case: a large Poissonian noise (k large) favors the emergence of periodic orbit.

Theorem 3 *For every fixed* $\alpha,\beta,k > 0$, *there exists* $\overline{\sigma}$ *such that the asynchronous* $\overline{p}$ *is linearly stable for every* $\sigma > \overline{\sigma}$.

This theorem shows that a large Brownian noise favors disorder in the spin distribution, though it is not enough to rule out existence of periodic solutions. Numerical simulations rather clearly suggest, however, that no periodic orbit should exist for large σ.

If we keep σ fixed (as well as α and β), and control both noise intensities by varying k, we appreciate the different effects of the two sources of noise on the synchronization. We have developed an heuristic argument using the approximation of $\tanh(\cdot)$ by its third order Taylor polynomial. Under this approximation, the infinite dymensional equation (15) can be reduced to a finite dimensional o.d.e., for the mean of of $\Sigma(t)$ and the first three moments of $\Lambda(t)$. For this reduced system the bifurcation analysis can be done explicitely. The results support the following conjecture for the "true" system (15).

Conjecture. Assume $\beta > 1$. Then there is $\tilde{\sigma} = \tilde{\sigma}(\alpha, \beta)$ such that:

1. For $\sigma \geq \tilde{\sigma}$ and any $k > 0$, (15) has a unique stationary solution, which is globally stable. In particular no non-constant periodic solution exists.
2. For $0 < \sigma < \tilde{\sigma}$ a non-constant periodic solution exists for k in two non-empty intervals $[k^-, k^- + \varepsilon]$ and $[k^+ - \varepsilon, k^+]$, with $0 < k^- < k^+ < +\infty$, ε small enough, and both k^+, k^- may depend on α, β and σ.

Moreover, for $\beta \leq 1$, (15) has a unique, globally stable stationary solution, for every $\alpha > 0, \sigma \geq 0$.

Periodic solutions "appear" at k^- and "disappear" at k^+ through Hopf bifurcations. The emergence of other patterns in the interval $[k^- + \varepsilon, k^+ - \varepsilon]$ is currently under investigation.

References

1. Acebrón, J.A., López Bonilla, L., Pérez Vicente, C.J., Ritort, F., Spigler, R.: The kuramoto model: A simple paradigm for synchronization phenomena. Reviews of modern physics **77**(1), 137 (2005)
2. Bertini, L., Giacomin, G., Pakdaman, K.: Dynamical aspects of mean field plane rotators and the kuramoto model. J. Statist. Phys. **138**, 270–290 (2010)
3. Brunel, N., Hakim, V.: Fast global oscillations in networks of integrate-and-fire neurons with low firing rates. Neural computation **11**(7), 1621–1671 (1999)
4. Carletti, T., Villari, G.: A note on existence and uniqueness of limit cycles for Liénard systems. Journal of mathematical analysis and applications **307**(2), 763–773 (2005)
5. Dai Pra, P., Fischer, M., Regoli, D.: A Curie-Weiss model with dissipation. Journal of Statistical Physics **152**, 37–53 (2013)
6. Dai Pra, P., Giacomin, G., Regoli, D.: Periodic behavior in a Curie-Weiss model with noisy rates. In preparation (2013)
7. Elowitz, M.B., Leibler, S.: A synthetic oscillatory network of transcriptional regulators. Nature **403**(6767), 335–338 (2000)
8. Garcia-Ojalvo, J., Elowitz, M.B., Strogatz, S.H.: Modeling a synthetic multicellular clock: repressilators coupled by quorum sensing. Proceedings of the National Academy of Sciences of the United States of America **101**(30), 10955–10960 (2004)
9. Giacomin, G., Pakdaman, K., Pellegrin, X., Poquet, C.: Transitions in Active Rotator Systems: Invariant Hyperbolic Manifold Approach. SIAM J. Math. Anal. **44**(6), 4165–4194 (2012)
10. Kuramoto, Y.: Chemical oscillations, waves, and turbulence. Courier Dover Publications, New York (2003)
11. McMillen, D., Kopell, N., Hasty, J., Collins, J.J.: Synchronizing genetic relaxation oscillators by intercell signaling. Proceedings of the National Academy of Sciences **99**(2), 679–684 (2002)
12. Pakdaman, K., Perthame, B., Salort, D.: Relaxation and self-sustained oscillations in the time elapsed neuron network model. SIAM J. Appl. Math. **73**(3), 1260–1279 (2013)

13. Pakdaman, K., Perthame, B., Salort, D.: Dynamics of a structured neuron population. Nonlinearity **23**(1), 55–75 (2010)
14. Romanczuk, P., Bär, M., Ebeling, W., Lindner, B., Schimansky-Geier, L.: Active brownian particles. The European Physical Journal Special Topics **202**(1), 1–162 (2012)
15. Sabatini, M., Villari, G.: Limit cycle uniqueness for a class of planar dynamical systems. Applied mathematics letters **19**(11), 1180–1184 (2006)
16. Sakaguchi, H., Shinomoto, S., Kuramoto, Y.: Phase transitions and their bifurcation analysis in a large population of active rotators with mean-field coupling. Progress of Theoretical Physics **79**(3), 600–607 (1988)
17. Scheutzow, M.: Periodic behavior of the stochastic brusselator in the mean-field limit. Probability theory and related fields **72**(3), 425–462 (1986)
18. Schweitzer, F.: Brownian agents and active particles. Collective dynamics in the natural and social sciences, With a foreword by J. Doyne Farmer. Springer Series in Synergetics. Springer-Verlag, Berlin Heidelberg New York (2003)
19. Shinomoto, S., Kuramoto, Y.: Phase transitions in active rotator systems. Progress of Theoretical Physics **75**(5), 1105–1110 (1986)
20. Touboul, J., Hermann, G., Faugeras, O.: Noise-induced behaviors in neural mean field dynamics. SIAM J. Applied Dynamical Systems **11**(1) 49–81 (2011)

Kinetic Equations and Stochastic Game Theory for Social Systems

Andrea Tosin

Abstract In this paper we present mathematical tools inspired by the kinetic theory, which can be used to model the social behaviors of large communities of individuals. The focus is especially on human societies, such as the population of a certain country, and on the interplays between concurrent social dynamics, for instance economic issues linked to the formation of political opinions, which sometimes can even degenerate into dramatic extreme events with massive impact (*Black Swans*). Starting from Boltzmann-type models, we present an evolution of the classical approach of statistical mechanics, whose hallmark is the use of stochastic game theory for the description of social interactions. By this we mean that the latter are modeled as games whose payoffs, however, are known only in probability. This is consistent with the basic unpredictability of human reactions, which ultimately cannot be compared to deterministic mechanical-like "collisions".

1 Introduction

Human societies are definitely complex systems, which modern applied mathematics is expected to help unravel [1]. The modeling of human behaviors appears indeed to be one of the next challenges of the applied mathematical research [29]. Thanks to its innate reductionism, mathematics can succeed in shedding some light on those intricate decision-based mechanisms which lead people to produce, mostly unconsciously, complex collective trends out of relatively elementary individual interactions.

A. Tosin (✉)
Istituto per le Applicazioni del Calcolo "M. Picone", Consiglio Nazionale delle Ricerche, via dei Taurini 19, 00185 Rome, Italy
e-mail: a.tosin@iac.cnr.it

A. Celletti, U. Locatelli, T. Ruggeri and E. Strickland (eds.): *Mathematical Models and Methods for Planet Earth.* Springer INdAM Series 6, DOI 10.1007/978-3-319-02657-2_4,

Several complexity issues, however, need to be addressed, which make the mathematical approach to social systems quite different from that to more classical physical systems (such as e.g., fluids or gases).

First of all, as already mentioned, large scale collective behaviors, which are ultimately the most interesting ones to predict, emerge spontaneously from interactions among few individuals at a small scale. This phenomenon is known as *self-organization* [26]. Collective behaviors can be sometimes quite irrational despite the fact that individual interactions follow relatively rational rules. The reason is that each individual is normally not even aware of the group s/he belongs to and of the group behavior s/he is contributing to, because s/he acts only locally [2]. Consequently, no individual has full access to group behaviors or can voluntarily produce and control them. Therefore, models are required to adopt *multiscale approaches* able to retain the proper amount of localized individual interactions, often expressed over networks [24], also within a collective description.

Secondly, individual interaction rules can be interpreted deterministically only up to a certain extent, due to the ultimate unpredictability of human reactions. It is the so-called *bounded rationality* [7,27], which makes two individuals react possibly not the same, even if they face the same conditions. In opinion formation problems this issue is of paramount importance, for the volatility of human behaviors can play a major role in causing rare and extreme events with massive impact, known under the evocative name of *Black Swans* in the socio-economic sciences after the celebrated book [31]. Mathematical models should be able to handle, within the aforesaid multiscale perspective, such *stochastic effects* at the level of individual interactions. In many cases, it is questionable whether they can or cannot be schematized as standard white noises.

Several other hallmarks of social systems, viewed as living complex systems, might be listed. For a more comprehensive coverage we refer interested readers to the recent literature [6, 11, 12, 14]. Here we confine the attention to the two issues set forth above, showing that a suitable evolution of the classical "collisional" kinetic approach can be a promising way of tackling them.

The kinetic representation is a *mesoscopic* one, meaning that it provides an ensemble statistical description of the particle system while grounding the evolutionary dynamics on small scale interactions among the individuals. As such, it is a possible option for dealing with the previously mentioned multiscale interplay between individuality and collectivity. Nevertheless, collisions among classical particles of inert matter are mostly ruled by deterministic mechanical principles, such as the balance of linear momentum and energy, which are not directly available in the case of interactions among living particles. In fact the latter are able to actively develop decision-based behavioral strategies, which, as recalled above, are neither fully deterministic nor actually mechanical. Therefore it is necessary to duly evolve the usual collisional framework so as to include forms of stochasticity in one-to-one interactions.

It is worth remarking, however, that unlike inert matter, whose mathematical description can be often grounded on consolidated physical theories, living matter still lacks a precise treatment in terms of quantitative theories. Hence, apart from the mathematical framework, the detailed modeling of small scale dynamics contains inevitably some heuristics. If, on the one hand, this can be a handicap for the predictive reliability of the models, hence for their immediate "industrial" use, on the other hand it offers mathematics the great opportunity to play a leading role in opening new ways of scientific investigation. Mathematical models can indeed fill the quantitative gap by acting themselves as paradigms for exploring and testing conjectures about the inner dynamics of systems. The latter may be much simpler than the observable large-scale outcomes and nevertheless much more difficult to isolate and ascertain phenomenologically. In doing so, it can happen that models put in evidence also facts not yet empirically observed, whereby scientists can be motivated to perform new specific experiments aiming at confirming or rejecting models' conjectures.

Finally, mathematics itself can take advantage of these applications for developing new mathematical methods and theories. In fact, nonstandard applications typically generate challenging analytical problems, whereby the role of mathematical research as a preliminary necessary step for mastering new models also at an industrial level is enhanced.

Coming to the structure of this paper, Section 2 introduces the basic ideas of the kinetic approach applied to social systems, which are then formalized in Boltzmann-like mathematical equations in Section 3. Next, Section 4 presents the evolution of such mathematical structures under the perspective of stochastic game theory. The latter proves to be a valid complement to the standard kinetic approach for modeling the intrinsic randomness of human behaviors. Interestingly, the new class of kinetic equations thus obtained contains classical "collisional" equations as a particular case. Section 5 discusses then a few meaningful variations of the mathematical framework suitable to capture further hallmarks of living social systems, which make the latter different from particles of inert matter. These variations include the so-called nonlinearly additive interactions, which essentially refer to the influence of the milieu on the individual decision-based dynamics, and the use of discrete microscopic states of the particles. Finally, Section 6 presents an example of application concerning the interplay between wealth redistribution policies and the evolution of the mass opinion about the doings of a certain government.

2 The kinetic approach

Social systems can be mathematically schematized as systems of particles, viz. individuals, which interact with one another thereby changing in time a certain state, which defines their social behavior. We generically call this state *activity* and denote it by $u \in U$, $U \subseteq \mathbb{R}^n$ being the activity domain. Particles with state u are also called *active particles*.

Often u is a scalar variable, i.e., $n = 1$. For example, it can represent the *opinion* of the individuals about a certain issue and its domain can be an interval such as $U = [-1, 1]$, where $u = -1$ stands for the strongest disagreement and $u = 1$ for the maximum agreement. Alternatively, if u is a political opinion, or an intended vote, then $u = \pm 1$ can represent the inclination toward either extreme wing of the political scenario. In socio-economic problems, u can instead denote the *social class* of the individuals. In this case, the domain U can be either a subset of the positive real axis, if u is intended to represent the material wealth, or again the same interval as before, if u is used more in the abstract as an indicator of poverty ($u = -1$) or wealthiness ($u = 1$).

When u is a vector variable, i.e., $n > 1$, the social behavior of the active particles is determined by various interconnected aspects accounted for by the components of u. For instance, the support or opposition to a government by the population of a country can be intimately related to the economic issues of wealth redistribution among the social classes. In this case, u can be taken as a two-dimensional variable, the first component denoting the level of well-being of the individuals and the second one their opinion about government's doings (cf. Sect. 6). More in general, a vector-valued activity can be handled by means of a decomposition of the system in *functional subsystems*, a topic that we will not specifically treat here. For a deep analysis we refer interested readers to [4–6].

The primary goal of models of social systems is to describe how the distribution of the activity within the considered population of active particles changes in time, in order to predict the evolution over time of social scenarios. To this purpose, in the kinetic approach one introduces the so-called *one-particle distribution function*

$$f = f(t, u) : [0, +\infty) \times U \to \mathbb{R}_+,$$

such that $f(t, u)\,du$ gives the (infinitesimal) number of active particles whose activity at time t belongs to the (infinitesimal) volume du centered at u in the space U of microscopic states. Under suitable integrability assumptions, namely $f(t) \in L^1(U)$ for all $t > 0$,

$$N_V(t) := \int_V f(t, u)\,du, \qquad V \subseteq U$$

is the number of active particles whose activity at time t belongs to a certain Borel-measurable subset V of the state space. Hence $N_U(t)$ is the total number of particles of the system at time t. If the latter is conserved in time, then f can be normalized with respect to it thereby becoming a probability density. In this case, informative statistics concerning the activity distribution can be computed as moments of the distribution function f, under the necessary further integrability assumptions.

A kinetic model consists in an evolution equation for the distribution function f derived consistently with the aforesaid meaning of the latter. In particular, the model has to formalize the concept of *interactions* among active particles, linking them to the time variation of f. A quite general conceptual scheme that can be adopted is the following:

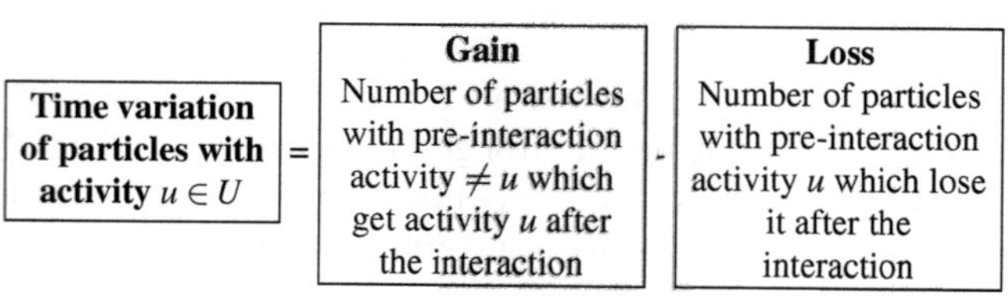

the gain and loss of particles with state u being evaluated in the unit time in terms of the statistical information provided by f. In particular, it is useful to recall two fundamental assumptions underlying the modeling of interactions in most kinetic models:

(H1) Only *pairwise interactions* are considered, meaning that interactions involving more than two active particles at once are disregarded as higher order effects.

(H2) The *factorization* of the *two-particle distribution function* f_2 is assumed:

$$f_2(t, u_1, u_2) = f(t, u_1)f(t, u_2)$$

in order to come to expressions of the gain and loss terms which only require the knowledge of f. This amounts to neglecting two-particle correlations when estimating the pairs of particles involved in the interactions, a fact that one has to accept as an approximation, however useful, of physical reality.

In the next sections we will present a few kinetic equations derived within this framework. They evolve from classical Boltzmann-type to more recent mathematical structures motivated by the quest for models able to include the specific complexity features discussed in the Introduction.

3 Boltzmann-Type models

Let us assume that pairwise interactions among active particles are *deterministic*, i.e., that there exists a mapping $\mathscr{C} : U^2 \to U^2$ such that $(u, v) = \mathscr{C}(u_*, v_*)$ is the post-interaction pair of activities corresponding to the pre-interaction pair (u_*, v_*). In particular, we require $\mathscr{C}$ to be a change of variable, i.e., one-to-one and onto. Then Boltzmann-type kinetic equations write, in the present context, as

$$\frac{\partial f}{\partial t} = Q[f, f], \tag{1}$$

where Q is the (bilinear) *interaction operator* expressing the balance between gain and loss of active particles with activity u:

$$\begin{aligned} Q[f, f](t, u) &= \int_U \left(\frac{1}{J_{\mathscr{C}}(u_*, v_*)} f(t, u_*)f(t, v_*) - f(t, u)f(t, v) \right) dv \\ &= \underbrace{\int_U \frac{1}{J_{\mathscr{C}}(u_*, v_*)} f(t, u_*)f(t, v_*)\,dv}_{\text{Gain}} - \underbrace{f(t, u) \int_U f(t, v)\,dv}_{\text{Loss}}. \end{aligned} \tag{2}$$

Models of social systems variously relying on the mathematical structure (1)–(2) can be found in, among others, [20,21,23,25,28,32], possibly with the inclusion of suitable *interaction kernels* (cf. Sect 5).

The pre-interaction activities u_*, v_* appearing in the gain term have to be understood as functions of the post-interaction ones u, v. In particular, they are those which generate the pair (u, v) after the interaction, namely $(u_*, v_*) = \mathscr{C}^{-1}(u, v)$. Integration with respect to v then counts all interactions which make one active particle shift to the activity u, independently of the activity earned by the companion particle (a similar argument holds also for the integration with respect to v in the loss term). Finally, $J_{\mathscr{C}}$ denotes the Jacobian of the transformation $\mathscr{C}$:

$$J_{\mathscr{C}}(u_*, v_*) := |\det \nabla \mathscr{C}(u_*, v_*)|,$$

which is needed for mass conservation purposes. By performing the change of variables $(u, v) = \mathscr{C}(u_*, v_*)$ in the gain term it can be checked that it results indeed

$$\int_U Q[f, f](t, u)\, du$$
$$= \int_U \int_U \left(\frac{1}{J_{\mathscr{C}}(u_*, v_*)} f(t, u_*) f(t, v_*) - f(t, u) f(t, v) \right) du\, dv = 0,$$

whereby integration over U in Equation (1) yields

$$\frac{d}{dt} \int_U f(t, u)\, du = 0.$$

Hence the total number N_U of active particles is constant in time. This property is often rephrased by saying that either the operator Q or the interactions that it models are conservative.

Remark 1 (Interactions as games) It is useful, also in view of the extensions that we will discuss in the next sections, to interpret pairwise interactions among active particles as two-player games. According to the terminology used in game theory, we can compare the pre-interaction activities u_*, v_* to the *game strategies* by which the two *players*, viz. the active particles, approach the *game*, namely the interaction. Then the post-interaction activities u, v are the *payoffs* of the game, that either player will use as new game strategy in future interactions. Within this analogy, the game described by Equation (1) is *evolutionary*, because the distribution of players' strategies evolves in time.

4 Stochastic game models

One of the most distinguishing complexity features of social systems recalled in the Introduction is the *bounded rationality*: the volatility of human behaviors makes it possible that individuals do not react in the same manner when facing the same situation. Then, recalling also Remark 1, we can draw a parallelism between the rationality of the behavior of active particles and the determinism of the game by which we model their interactions. A deterministic game, i.e., one featuring a deterministic relationship between pre- and post-interaction strategies, means a fully rational behavior. On the other hand, the concept of bounded rationality calls for game-type models of interactions in which the post-interaction strategies (payoffs) are known only in terms of a certain probability distribution. Therefore we use the term *stochastic games* for this type of interactions.

Stochastic game-type interactions can be implemented in Equation (1) by duly generalizing the structure of the interaction operator Q, particularly the gain term (cf. [8]):

$$Q[f,f](t,u) = \underbrace{\iiint_{U^3} \mathscr{P}((u_*,v_*) \to (u,v)) f(t,u_*) f(t,v_*)\, du_*\, dv_*\, dv}_{\text{Gain}}$$

$$\underbrace{- f(t,u) \int_U f(t,v)\, dv}_{\text{Loss}}. \tag{3}$$

In practice, the mapping $\mathscr{C}$ is replaced by the *transition probability density* $\mathscr{P}$, namely the probability density of the payoff (u,v) conditioned to the pre-interaction strategy pair (u_*,v_*). The following property must hold:

$$\iint_{U^2} \mathscr{P}((u_*,v_*) \to (u,v))\, du\, dv = 1 \quad \forall (u_*,v_*) \in U^2, \tag{4}$$

which ensures $\int_U Q[f,f](t,u)\, du = 0$, hence again the conservativeness of the interactions.

From Equation (3) the deterministic setting of Equation (2) can be recovered by assuming

$$\mathscr{P}((u_*,v_*) \to (u,v)) \propto \delta_{\mathscr{C}^{-1}(u,v)}(u_*,v_*), \tag{5a}$$

where δ is the Dirac's delta. Indeed this implies that, for a fixed post-interaction activity pair (u,v), the only pre-interaction activity pair which can produce it is $(u_*,v_*) = \mathscr{C}^{-1}(u,v)$ (notice that in Eq. (3) it is convenient to regard $\mathscr{P}$ as a function of u_*, v_* parameterized by u, v). Since $\mathscr{P}$ has to be a probability density with respect to (u,v), the correct scaling in Equation (5a) is actually

$$\mathscr{P}((u_*,v_*) \to (u,v)) = \frac{1}{J_{\mathscr{C}}(u_*,v_*)} \delta_{\mathscr{C}^{-1}(u,v)}(u_*,v_*), \tag{5b}$$

which guarantees the fulfillment of Equation (4).

An alternative form of the gain term in Equation (3) is obtained by defining the quantity

$$\mathscr{A}(u_* \to u | v_*) := \int_U \mathscr{P}((u_*,v_*) \to (u,v))\, dv, \tag{6}$$

which according to Equation (4) satisfies

$$\int_U \mathscr{A}(u_* \to u | v_*)\, du = 1 \quad \forall (u_*,v_*) \in U^2. \tag{7}$$

It represents directly the probability density of the payoff u accessible by the active particle with pre-interaction strategy u_*, given the pre-interaction strategy v_* of the other particle. The interaction operator rewrites then as

$$Q[f,f](t,u) = \iint_{U^2} \mathscr{A}(u_* \to u|v_*) f(t,u_*) f(t,v_*)\, du_*\, dv_* \\ - f(t,u) \int_U f(t,v)\, dv. \tag{8}$$

This new form of Q induces a useful terminology for identifying the particles which take part in the interactions along with their respective roles:

- The particle with activity u, namely the generic representative entity addressed by the operator Q, is the *test particle.*
- The particle with pre-interaction activity u_*, which can shift to u after the interaction, is a *candidate particle.*
- The particle with pre-interaction activity v_*, which triggers the change of activity of the candidate particle, is a *field particle.*

Modeling social interactions as stochastic games via Eq. (8) means therefore assuming that the candidate particle gets in probability the state of the test particle, and that the latter loses it, because of kinds of "generalized collisions" with field particles.

4.1 Types of games

Equations (1)–(3) (or (1)–(8)) provide a mathematical framework which can be used to generate specific models, up to detailing the transition probability density $\mathscr{P}$ (or $\mathscr{A}$). This corresponds to specifying the *social game* that the active particles play when they interact with one another. In order to classify some of the most popular types of games, it is useful to introduce a metric $d : U^2 \to \mathbb{R}_+$ in the activity space U, which measures the distance between the activities of the interacting particles. For $U \subseteq \mathbb{R}^n$, d will be most often the usual Euclidean distance $d(u,v) = |v-u|$. Then interactive games among active particles can be characterized by the relationship between the pre- and post-interaction activity distances, $d(u_*, v_*)$ and $d(u,v)$, respectively.

Cooperation or **consensus**

This game is such that

$$d(u,v) \leq d(u_*, v_*),$$

see Figure 1. Therefore active particles reduce their activity distance, experiencing a sort of attraction. If the activity is their wealth status, this corresponds to a cooperative attitude of wealthy social classes toward poor ones. If instead the activity is an opinion, this corresponds to the tendency to agree (consensus).

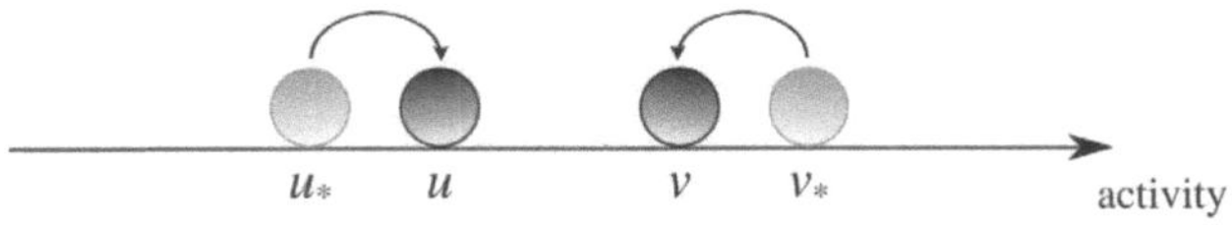

Fig. 1 Cooperating active particles reduce the distance between their activities

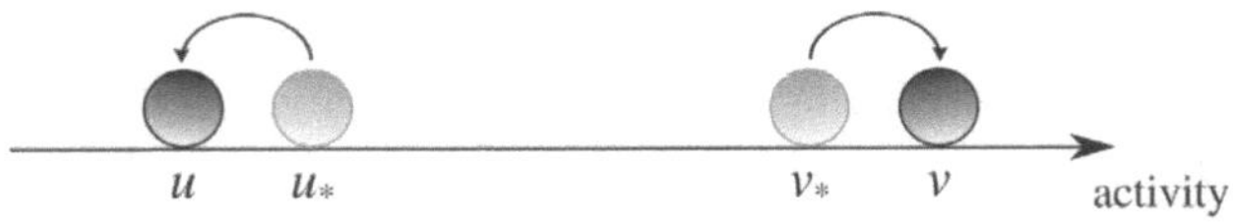

Fig. 2 Competing active particles increase the distance between their activities

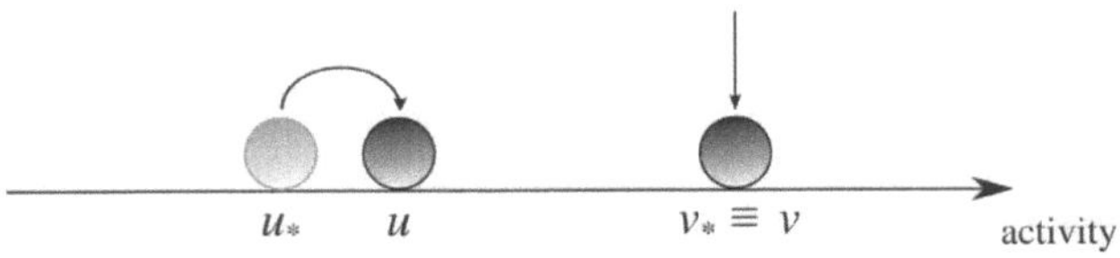

Fig. 3 An active particle that learns from another one approaches the activity of the latter

Competition or **dissent**

This game is such that

$$d(u, v) \geq d(u_*, v_*),$$

see Figure 2. Hence active particles increase their activity distance, experiencing a sort of repulsion. If the activity is their wealth status, this corresponds to a competitive attitude of wealthy social classes against poor ones. If instead the activity is an opinion, this corresponds to the tendency to disagree (dissent).

Learning

This game is such that

$$v = v_* \quad \text{and} \quad d(u, v_*) \leq d(u_*, v_*),$$

see Figure 3. Here the main point is that one of the interacting particles learns from the other, i.e., it approaches the activity of the latter. The second particle is instead assumed not to change activity during the interaction. For instance, if the activity is an opinion this may correspond to active particles open to persuasion.

Hiding-chasing

This game is such that

$$\begin{cases} d(u, v_*) \leq d(u_*, v_*) \\ d(u_*, v) \geq d(u_*, v_*). \end{cases}$$

see Figure 4. Here in principle both interacting particles change their activity but with opposite goals: One of them behaves so as to increase the distance (hiding),

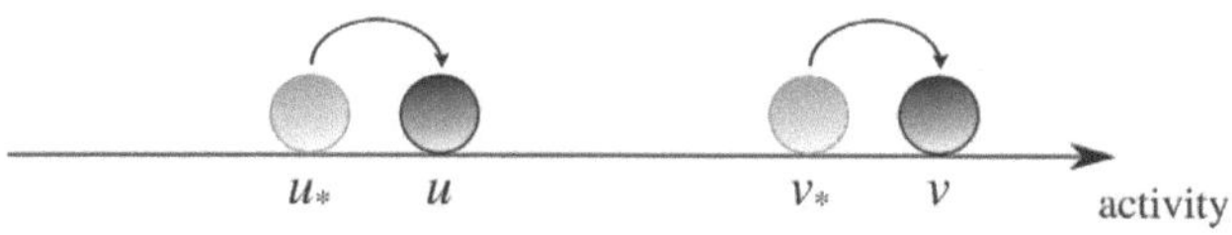

Fig. 4 A hiding-chasing game is like a pursuit between an escaping and a chasing particle

whereas the other aims at reducing it (chasing). If the activity is the wealth status of the particles, this may correspond to the case in which either wealthy social classes produce further wealth that also poor classes benefit from or, conversely, the whole society progressively loses purchasing power (because e.g., of a global financial crisis) so that both wealthy and poor social classes become poorer and poorer.

Remark 2 The relationships above between the pre- and post-interaction activity distances are given deterministically, which is in principle appropriate only when the transition probability density $\mathscr{P}$ is as in Equation (5b). In this case we also have $d(u, v) = d(\mathscr{C}(u_*, v_*))$. In general, for genuinely stochastic interactions, such relationships have to be thought of as referred to single realizations of the games. More precisely, one should use the metric d for defining a subset $V_{(u_*, v_*)} \subseteq U^2$ which, for a given pre-interaction activity pair (u_*, v_*), carries all the transition probability (cf. also Eq. (4)). For instance, in case of cooperation/consensus we define

$$V_{(u_*, v_*)} := \{(u, v) \in U^2 : d(u, v) \leq d(u_*, v_*)\}$$

and then we say that $\mathscr{P}((u_*, v_*) \to (\cdot, \cdot))$ is supported on $V_{(u_*, v_*)}$, i.e.,

$$\iint_{V_{(u_*, v_*)}} \mathscr{P}((u_*, v_*) \to (u, v))\,du\,dv = 1,$$

which implies that the $\mathscr{P}$-probability measure of the set $U^2 \setminus V_{(u_*, v_*)}$ is zero, namely that a transition from (u_*, v_*) to (u, v) with $d(u, v) > d(u_*, v_*)$ is statistically impossible.

The other three cases can be handled analogously.

More than one type of game can be played simultaneously by active particles. For instance, in [6, 13, 16] active particles cooperate or compete depending on their pre-interaction activity distance compared to a reference threshold γ. In those works the activity denotes the social class of the particles, the underlying idea being that two particles with close wealth states, i.e., such that $d(u_*, v_*) \leq \gamma$, may have conflicting interests whereas two particles with markedly dissimilar wealth states, i.e., such that $d(u_*, v_*) > \gamma$, may be forced to cooperate by an imposed social policy. Other examples of distance-dependent game switch can be found in [3] referred to applications such as electoral competition on the Internet and job mobility.

5 Further extensions of the stochastic game approach

Equations (3) or (8) assume a constant unit interaction frequency among pairs of active particles. An immediate extension, which has also a counterpart in the classical collisional kinetic theory, consists in including an *interaction rate*, say $\eta : U^2 \to \mathbb{R}_+$, which depends on the pre-interaction activities. Hence Equation (3) rewrites as

$$
\begin{aligned}
Q[f,f](t,u) = &\iiint_{U^3} \eta(u_*,v_*)\mathscr{P}((u_*,v_*) \to (u,v)) f(t,u_*) f(t,v_*)\, du_*\, dv_*\, dv \\
&- f(t,u) \int_U \eta(u,v) f(t,v)\, dv
\end{aligned}
\tag{9a}
$$

and analogously Equation (8) as

$$
\begin{aligned}
Q[f,f](t,u) = &\iint_{U^2} \eta(u_*,v_*)\mathscr{A}(u_* \to u|v_*) f(t,u_*) f(t,v_*)\, du_*\, dv_* \\
&- f(t,u) \int_U \eta(u,v) f(t,v)\, dv.
\end{aligned}
\tag{9b}
$$

Basically, η says how much frequent are interactions between two particles with respective activities u_*, v_*. For instance, depending on the specific application it might be argued that individuals with very different activities interact less frequently than those with more similar activities. Hence, for example, in some cases one may assume that η decreases with the distance $d(u_*, v_*)$, cf. e.g., [13]. The particular case of constant η can be reinterpreted as a model of a well mixed population, where each active particle equally interacts with any other, see [16].

Two other conceptually relevant evolutions of these mathematical structures are discussed in the next subsections.

5.1 *Nonlinearly additive interactions*

According to the form of the operator Q presented so far, interactions among pairs of active particles are in some sense *linear*. This can be understood e.g., from Equation (9b) by rewriting the gain term as

$$
\int_U \underbrace{\left(\int_U \eta(u_*,v_*)\mathscr{A}(u_* \to u|v_*) f(t,v_*)\, dv_* \right)}_{\substack{\text{Total action applied by field particles} \\ \text{on the candidate particle with activity } u_*}} f(t,u_*)\, du_*.
$$

Apart from the fact that the time variation of u_* cannot be ascribed, in general, to a deterministic microscopic rule, we see that the collective action of field particles *on a single candidate particle* is formally a mean field-like superposition of effects.

In particular, we notice that the interaction rules expressed by the transition probability density are parameterized only by the pre-interaction states of the interacting particles. This implies that the presence of other particles, namely the *milieu* where interactions take place, has no effect on the outcome of single pairwise interaction instances. If this can be reasonable for mechanical interactions among inert particles, it can be argued that social interactions among living particles can instead produce different outputs also starting from the same inputs, because human reactions are highly milieu-sensitive.

This discussion can be introduced in models (9a), (9b) by assuming that the transition probability densities $\mathscr{P}$, $\mathscr{A}$, respectively, depend also on the distribution function f. In fact the latter brings information about the collective state of the system, namely indeed the milieu, where interactions develop. For example, in [13] the effects of social cooperation and competition on the wealth redistribution are examined under the assumption that either type of interaction is triggered by a threshold γ, which depends on the instantaneous distribution of the social classes. In more detail, as already mentioned at the end of Section 4, it is assumed that cooperation takes place between two individuals with sufficiently different wealth state, thus when $d(u_*, v_*) > \gamma$, whereas competition occurs between two individuals with similar states, hence when $d(u_*, v_*) \leq \gamma$. The relevant point here is that the value of γ depends on the *social gap* S, namely the difference between the number of poor and wealthy individuals, in such a way that the higher S the higher γ, i.e., the higher the inclination of active particles to competition in an attempt to individually raise their own well-being. Using a scalar activity $u \in [-1, 1]$ and agreeing that $u < 0$, $u > 0$ identify poor and wealthy social classes, respectively, the instantaneous social gap can be computed as

$$S(t) = \int_{-1}^{0} f(t, u)\,du - \int_{0}^{1} f(t, u)\,du. \tag{10}$$

Therefore γ, as a function of S, depends functionally on f and such a dependence is transferred to the transition probability densities $\mathscr{P}$, $\mathscr{A}$ for, as explained above, the switch between cooperation and competition depends on γ.

When the transition probability densities depend on the distribution function we speak of *nonlinearly additive* interactions. Formally nothing changes in Eqs. (9a), (9b) but the notation, since the transition probability densities are now written as $\mathscr{P}[f]$, $\mathscr{A}[f]$ in order to stress their functional dependence on f. On the other hand, the gain term reads now

$$\int_U \left(\int_U \eta(u_*, v_*) \mathscr{A}[f](u_* \to u | v_*) f(t, v_*)\,dv_* \right) f(t, u_*)\,du_*,$$

whence we see that the total action in parenthesis applied on the candidate particle with activity u_* is still the sum of individual actions, each of which however depends also on the distribution of field particles other than that triggering the interaction. Ultimately, the sum of the field actions does not scale linearly with the distribution of field particles, whence the use of the expression "nonlinearly additive" to refer to interactions in this context. Some aspects of the qualitative analysis of Equation (1) with nonlinearly additive interactions are treated in the paper [9].

Another source of nonlinearity in the interactions is the interaction rate η, which for fixed u_* can be compactly supported in U meaning that the candidate particle does not necessarily interact with all field particles but just with a subset of them (localized interactions). For instance, particles may be able to interact only with a characteristic number of other individuals, say N_0, which corresponds to their limited capacity to retain and process the outer information. This fact can be expressed by assuming that $\eta(u_*, \cdot)$ is supported in the closed ball $\bar{B}_R(u_*) \subseteq U$ centered at u_* and whose radius R is adapted in such a way that the number of field particles contained in $\bar{B}_R(u_*)$ equals N_0. In formulas:

$$R = \inf\left\{r > 0 : \int_{\bar{B}_r(u_*)} f(t, v_*)\,dv_* \geq N_0\right\},$$

where "$\geq$" instead of "$=$" is necessary in order to guarantee that R be always well defined (indeed, depending on the distribution f, a ball centered at u_* and containing exactly N_0 particles might not exist). Consequently, η depends in turn on f because $\operatorname{supp}\eta(u_*, \cdot) = \bar{B}_R(u_*)$ and R is defined via f. In order to stress this fact we write $\eta[f]$ rather than simply η.

Interactions of the type set forth above are often called *topological*. They have been introduced in [10] with reference to the dynamics of swarms of birds and then formalized mathematically in kinetic or measure-based models in [15,22]. Opposed to them are *metric* interactions, which are still characterized by a compactly supported interaction rate $\eta(u_*, \cdot)$ but the size of the support is fixed *a priori* on the basis of purely metric arguments, for instance the fact that particles interact only with close neighbors (in the sense of the distance in U). By definition, metric interactions do not depend on the distribution of field particles, hence they actually do not induce any further nonlinearity in the interaction operator Q.

5.2 Discrete state models

A social state is sometimes better individuated by means of *classes* (or ranges) of values rather than by a continuous variable. This is particularly true when it refers to originally non-numerical quantities, such as e.g., opinions or social classes, which are mapped to numerical values amenable to mathematical analysis. Most of the examples cited so far are actually of this type, to which we further add [17–19]. It is therefore of some interest to find the appropriate formulation of Equation (1), and especially of the interaction operator Q, when the activity domain U is a lattice, namely

$$U = \{u_i\}_{i \in I} \subset \mathbb{R}^n, \quad I \subseteq \mathbb{Z}^n.$$

Each discrete value u_i of the activity is called an *activity class*, i being an integer multi-index. Parallelly, the kinetic function f is written as an atomic distribution over such a lattice with time-varying coefficients:

$$f(t, u) = \sum_{i \in I} f_i(t)\delta_{u_i}(u). \tag{11}$$

In particular, $f_i(t)$ is the number of active particles which at time t are in the activity class u_i. In the following we will also use the vector notation $\boldsymbol{f} := (f_i)_{i\in I}$. Also $\mathscr{P}$ and $\mathscr{A}$, as probability densities over the payoffs u, v, take an atomic form, for now payoffs belong to the above activity lattice:

$$\begin{aligned}\mathscr{P}((u_*,v_*)\to(u,v)) &= \sum_{i,j\in I}\mathscr{P}^{ij}(u_*,v_*)\delta_{u_i}(u)\otimes\delta_{u_j}(v),\\ \mathscr{A}(u_*\to u|v_*) &= \sum_{i\in I}\mathscr{A}^i(u_*,v_*)\delta_{u_i}(u).\end{aligned} \tag{12}$$

In view of Equation (6) it results $\mathscr{A}^i = \sum_{j\in I}\mathscr{P}^{ij}$, with moreover

$$\sum_{i,j\in I}\mathscr{P}^{ij}(u_*,v_*) = \sum_{i\in I}\mathscr{A}^i(u_*,v_*) = 1 \quad \forall(u_*,v_*)\in U^2,$$

cf. Equations (4), (7).

In order to come to an evolution equation for the special distribution function (11) we refer to the weak form of the kinetic equation (1), in which we read $f(t,\cdot)$ and $\mathscr{A}(u_*\to\cdot|v_*)$ conveniently as finite measures on U. Namely:

$$\begin{aligned}\frac{d}{dt}\int_U\varphi(u)f(t,du) &= \int_U\varphi(u)Q[f,f](t,u)\,du\\ &= \iint_{U^2}\eta(u_*,v_*)\left(\int_U\varphi(u)\mathscr{A}(u_*\to du|v_*)\right)f(t,du_*)\otimes f(t,dv_*)\\ &\quad -\iint_{U^2}\varphi(u)\eta(u,v)f(t,du)\otimes f(t,dv) \qquad \forall\varphi\in C_c^\infty(U),\end{aligned}$$

$C_c^\infty(U)$ being the space of infinitely differentiable test functions with compact support in U. Here we have used the form (9b) of the interaction operator Q with the transition probability density $\mathscr{A}$. Technical calculations using the form (9a) with $\mathscr{P}$ are completely analogous, thus left as an exercise to the reader.

The weak form above is suited to the distributions (11), (12), which once plugged into produce

$$\sum_{i\in I}\left(\frac{df_i}{dt}-\sum_{h,k\in I}\eta(u_h,u_k)\mathscr{A}^i(u_h,u_k)f_hf_k+\sum_{k\in I}\eta(u_i,u_k)f_kf_i\right)\varphi(u_i)=0$$

for all $\varphi\in C_c^\infty(U)$. Setting

$$\eta_{hk} := \eta(u_h,u_k), \quad \mathscr{A}^i_{hk} := \mathscr{A}^i(u_h,u_k)$$

and invoking the arbitrariness of the test function we are finally led to

$$\frac{df_i}{dt} = \sum_{h,k\in I}\eta_{hk}\mathscr{A}^i_{hk}f_hf_k - f_i\sum_{k\in I}\eta_{ik}f_k, \qquad i\in I, \tag{13a}$$

namely a system of ordinary differential equations whose unknown is actually the vector $\boldsymbol{f} = (f_i)_{i\in I}$. In case of nonlinearly additive interactions the equation above is formally rewritten as

$$\frac{df_i}{dt} = \sum_{h,k\in I} \eta_{hk}[\boldsymbol{f}]\mathscr{A}^i_{hk}[\boldsymbol{f}]f_hf_k - f_i\sum_{k\in I}\eta_{ik}[\boldsymbol{f}]f_k, \qquad i \in I, \tag{13b}$$

so as to stress the dependence of the interaction rate and the transition probabilities on the discrete distribution function $\boldsymbol{f}$.

For fixed $(h,k) \in I^2$, $\{\mathscr{A}^i_{hk}\}_{i\in I}$ is a discrete probability distribution over the payoff u_i, indeed $\sum_{i\in I}\mathscr{A}^i_{hk} = 1$. This guarantees, once again, the conservativeness of the interactions. In fact summing both sides of either Equation (13a) or Equation (13b) over i yields $\frac{d}{dt}\sum_{i\in I} f_i(t) = 0$, that is, taking into account Equation (11), $\frac{d}{dt}\int_U f(t,du) = 0$.

Due to the discrete activity setting, $\mathscr{A}^i_{hk}$ is directly the probability of the transition from the candidate state u_h to the test state u_i triggered by the field state u_k. In this context, $\{\mathscr{A}^i_{hk}\}_{h,k,i\in I}$ is often called the *table of games* as it encodes the rules of the game played by active particles.

6 An example of application to socio-economic dynamics

We now exemplify the use of the mathematical structures discussed so far, in particular Equation (13b), for addressing a problem of socio-economic dynamics which has been extensively investigated in [6, 13].

The context is that of support or opposition to the economic policy of a government expressed by the population of a country depending on the individual and collective level of well-being. Therefore the active particles are the individuals of the population with a vector-valued discrete activity $u_i = (u^1_{i_1}, u^2_{i_2}) \in \mathbb{R}^2$, the first component representing their wealth state (social class) and the second one their opinion about the government policy.

In more detail, we identify the following structured lattice of social classes and political opinions:

$$U = \left\{\frac{2}{n-1}i_1 - \frac{n+1}{n-1}\right\}_{i_1=1,\dots,n} \times \left\{\frac{2}{m-1}i_2 - \frac{m+1}{m-1}\right\}_{i_2=1,\dots,m},$$

namely the Cartesian product of two uniformly spaced grids in the interval $[-1, 1]$ with respective grid steps $\Delta u^1 = \frac{2}{n-1}$ and $\Delta u^2 = \frac{2}{m-1}$:

$$u^1_1 = -1, \dots, u^1_{\frac{n+1}{2}} = 0, \dots, u^1_n = 1, \qquad u^2_1 = -1, \dots, u^2_{\frac{m+1}{2}} = 0, \dots, u^2_m = 1.$$

Since we have $n > 1$ social classes and $m > 1$ opinion classes the domain of the multi-index $i = (i_1, i_2)$ is $I = \{1, \dots, n\} \times \{1, \dots m\}$. We agree that $u^1_{i_1} < 0$ represents the poor social classes and $u^1_{i_1} \geq 0$ the wealthy ones, and likewise $u^2_{i_2} < 0$ stands for opposition and $u^2_{i_2} \geq 0$ for support to the government doings.

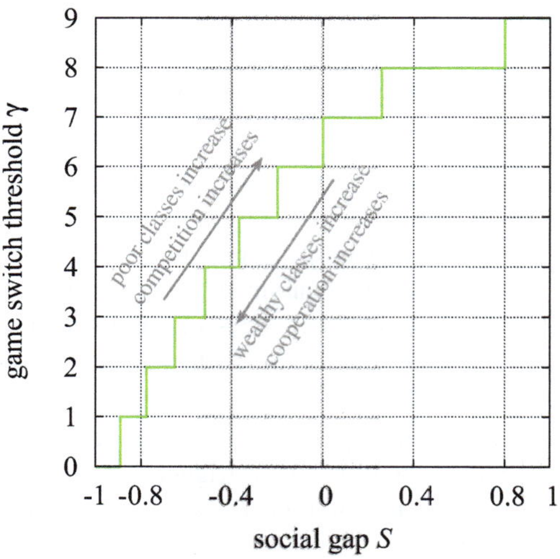

Fig. 5 The threshold γ between cooperation and competition as a function of the social gap S with $n = 9$ social classes

Wealth redistribution is modeled as a cooperative/competitive game among active particles as described in Section 4.1. In particular, the switch between either type of game is regulated by a threshold γ as discussed in Section 5.1, which depends on the social gap S:

$$S(t) = \underbrace{\sum_{i_1=1}^{\frac{n-1}{2}} \sum_{i_2=1}^{m} f_i(t)}_{\substack{\text{total individuals}\\ \text{in poor social classes}}} - \underbrace{\sum_{i_1=\frac{n+3}{2}}^{n} \sum_{i_2=1}^{m} f_i(t)}_{\substack{\text{total individuals}\\ \text{in wealthy social classes}}},$$

cf. Equation (10). We recall that if the activity distance is under threshold active particles compete, whereas if it is above threshold they cooperate. The graph in Figure 5 assumes then that in a society with predominance of wealthy social classes ($S < 0$) individuals tend to cooperate (low γ) and, conversely, that competition gets stronger and stronger (high γ) as the number of poor social classes increases ($S > 0$).

Alternatively one can assume that γ is constant, which means that the level of cooperation or competition is independent of the socio-economic dynamics. Interestingly, this can be interpreted as the fact that the government maintains the full control over the wealth redistribution dynamics, as opposed to the previous case in which they are instead left more freely to the market.

Opinion dynamics about the government policy are instead modeled mainly as a self-conviction. The idea is that poor individuals in a poor society tend in particular to distrust the government, while wealthy individuals in a wealthy society tend to trust it. On the other hand, poor individuals in a wealthy society and wealthy individuals in a poor society feature the most uncertain behavior. In fact the former may either distrust the government, in spite of the collective wealthiness, because of their low social condition or trust it, in spite of their poverty, because the collective affluence gives them chances to improve their condition. Analogously, the latter may either distrust or trust the government because of converse arguments. The mean social class:

$$\sum_{i\in I} u^1_{i_1} f_i(t) \tag{14}$$

is assumed as an indicator of the global economic condition of the society, namely it is used to decide whether a society is poor (when it is negative) or wealthy (when it is positive) on average.

Technically, the table of games $\mathscr{A}^i_{hk}[\boldsymbol{f}]$ is modeled as the product of two factors corresponding to wealth redistribution and opinion dynamics, respectively:

$$\mathscr{A}^i_{hk}[\boldsymbol{f}] = (\mathscr{A}')^{i_1}_{h_1k_1}[\boldsymbol{f}] \times (\mathscr{A}'')^{i_2}_{h}[\boldsymbol{f}].$$

The functional dependence of either factor on $\boldsymbol{f}$ includes the influence of the social gap (in $\mathscr{A}'$) and of the mean social class (in $\mathscr{A}''$). By a careful modeling of the cooperative/competitive interactions underlying wealth redistribution dynamics, namely the factor $\mathscr{A}'$, it is possible to guarantee that the mean social class (14) is conserved in time, a fact that we assume here. For the detailed expression of the table of games we refer interested readers to the original paper [13].

Figure 6 shows some asymptotic scenarios predicted by the model starting from a homogeneous distribution of social classes and political opinions in a society which is economically "neutral" on average (i.e., the mean social class is zero). Two test cases are considered corresponding to as many different values of the threshold γ, which is kept constant so as to simulate a strong governance on the economic policy. In both cases the model predicts the emergence of a group of interest formed by wealthy social classes, which markedly trust the government policy. On the other hand, poor social classes manifest a more varied political opinion encompassing all opinion classes, however with a tendency to trust the government especially in the context of a mainly cooperative economic policy (i.e., $\gamma = 3$).

Figure 7 refers instead to the case of a society which is poor on average (the mean social class is -0.4) and considers both a constant and a variable threshold γ ruling the switch between the cooperative and competitive attitude of the individuals. For strong governance (constant γ) the asymptotic scenarios are qualitatively opposite to those observed in Figure 6. The group of interest which now forms encompasses poor social classes, which tend to markedly distrust the government policy. This can be interpreted as an effect of the general lack of confidence due to the depressed economic condition of the society. However, if the imposed policy is mainly cooperative ($\gamma = 3$) then middle classes are still present, which express at least a mild support to the government. Conversely, if the policy is highly competitive ($\gamma = 7$) then middle classes disappear and only very wealthy ones, with quite mixed-up political opinions, remain to counterbalance the even stronger radicalization to opposition of poor classes.

Interestingly, the radicalization of the opposition predicted by the model is even more stressed under a weak governance (variable γ), because wealthy social classes

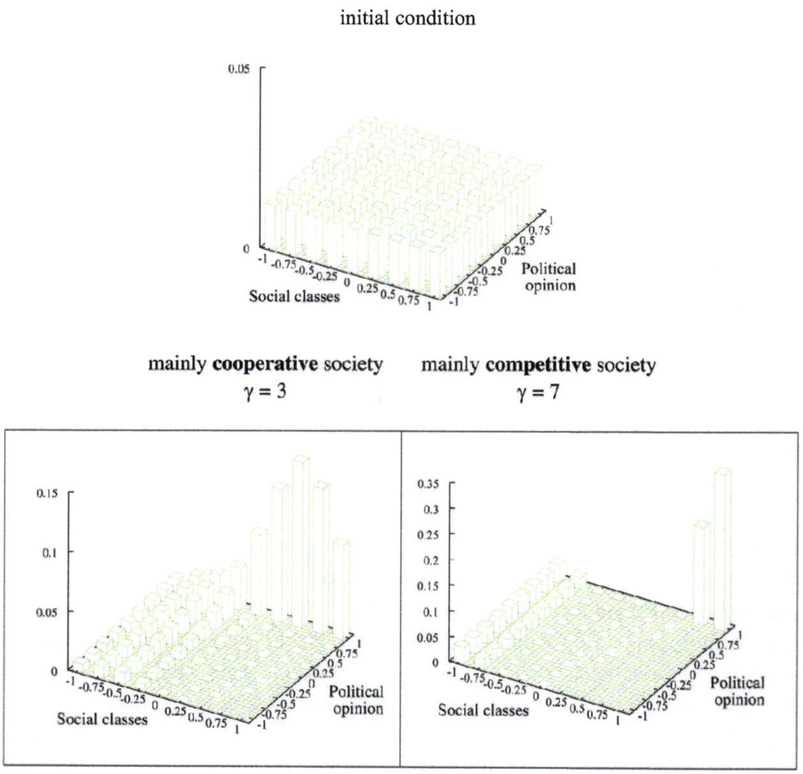

Fig. 6 Asymptotic scenarios of the distribution of social classes and political opinions in a society which is "neutral" on average (i.e., mean social class 0), under either a mainly cooperative ($\gamma = 3$) or a mainly competitive ($\gamma = 7$) attitude of the individuals. A lattice with $n = m = 9$ social and opinion classes has been used

may succeed in imposing large competition gaps for preserving their own well-being. Then the society tends to split almost completely into the lowest class, which is markedly against the government, and the highest class, which is instead innerly torn as far as the support to the government policy is concerned. This situation can prelude extreme events, such as popular rebellions, which could be unpredictable simply on the basis of the supposed social trend under a stronger governance. Hence model insights might serve as premonitory signals for the appearance of the so-called *Black Swans* [31].

In order to bring the latter observation to a quantitative level, we can choose to look at the time evolution of the distance between the distribution $\boldsymbol{f}$ under strong governance and the corresponding distribution under weak governance, starting from the same value of the threshold γ. This amounts to computing, for instance,

$$\max_{i_2=1,\dots,m} \sum_{i_1=1}^{n} \left| f_i^{\text{strong}}(t) - f_i^{\text{weak}}(t) \right|,$$

which is a particular norm in $\mathbb{R}^2$ of the difference $f^{\text{strong}}(t) - f^{\text{weak}}(t)$. Of course, also a different norm can be used since in finite dimension all norms are equivalent. Figure 8 shows the time trend of this quantity for both $\gamma = 3$ and $\gamma = 7$ referred to the test cases illustrated in Figure 7. It is interesting to notice the singular points in both curves, morally comparable to *tipping points* typical of catastrophe theory, which denote an abrupt turnround with respect to the possibly expected decrease to zero of the distance between the two distributions. These singularities confirm that at some point the social trend under weak economic governance departs from that under strong governance, possibly preluding crucial radicalizations.

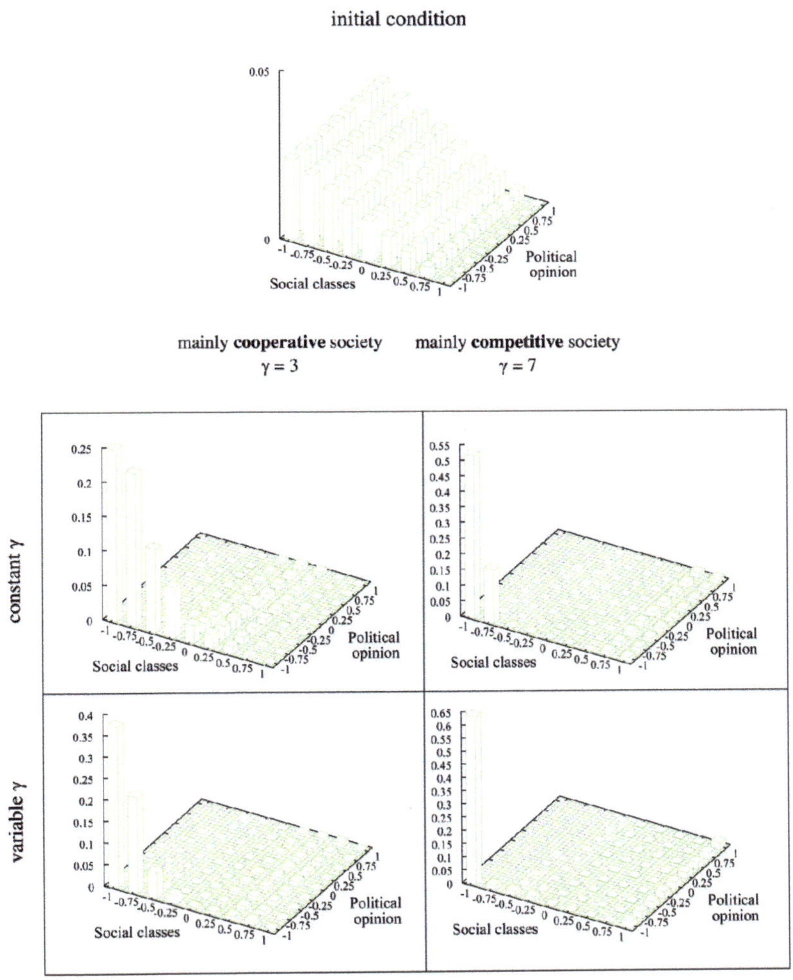

Fig. 7 Asymptotic scenarios of the distribution of social classes and political opinions in a society which is poor on average (mean social class -0.4) under both constant and variable threshold γ, starting from either a mainly cooperative ($\gamma = 3$) or a mainly competitive ($\gamma = 7$) attitude of the individuals. A lattice with $n = m = 9$ social and opinion classes has been used

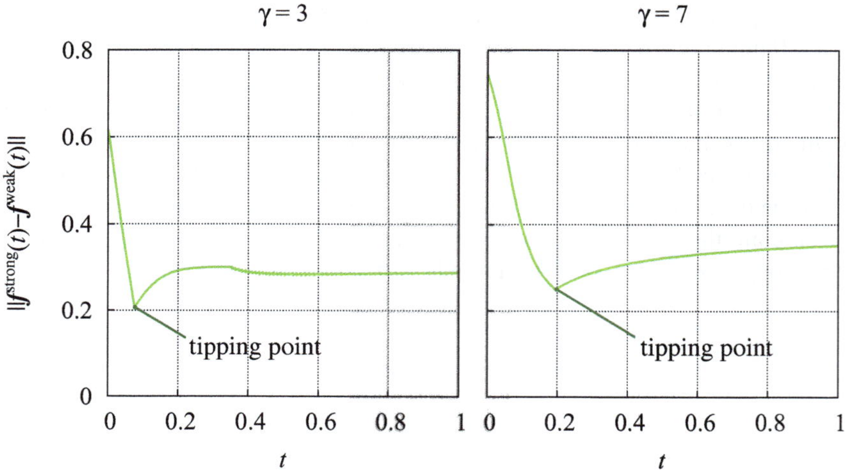

Fig. 8 Synthetic comparison over time of the social trends under strong and weak economic governance

References

1. Unraveling Complex Systems (2011). URL http://www.mathaware.org/mam/2011. Mathematics Awareness Month of the American Mathematical Society, the American Statistical Association, the Mathematical Association of America, and the Society for Industrial and Applied Mathematics
2. Agrawal, A., Kapur, D., McHale, J.: How do spatial and social proximity influence knowledge flows? Evidence from patent data. J. Urban Econ. **64**(2), 258–269 (2008)
3. Ajmone Marsan, G.: New paradigms towards the modelling of complex systems in behavioral economics. Math. Comput. Modelling **50**(3-4), 584–597 (2009)
4. Ajmone Marsan, G.: On the modelling and simulation of the competition for a secession under media influence by active particles methods and functional subsystems decomposition. Comput. Math. Appl. **57**(5), 710–728 (2009)
5. Ajmone Marsan, G., Bellomo, N., Egidi, M.: Towards a mathematical theory of complex socio-economical systems by functional subsystems representation. Kinet. Relat. Models **1**(2), 249–278 (2008)
6. Ajmone Marsan, G., Bellomo, N., Tosin, A.: Complex Systems and Society – Modeling and Simulation. SpringerBriefs in Mathematics. Springer, New York (2013)
7. Ariel, R.: Modeling Bounded Rationality. MIT Press, Cambridge, MA (1998)
8. Arlotti, L., Bellomo, N., De Angelis, E.: Generalized kinetic (Boltzmann) models: mathematical structures and applications. Math. Models Methods Appl. Sci. **12**(4), 567–591 (2002)
9. Arlotti, L., De Angelis, E., Fermo, L., Lachowicz, M., Bellomo, N.: On a class of integro-differential equations modeling complex systems with nonlinear interactions. Appl. Math. Lett. **25**(3), 490–495 (2012)
10. Ballerini, M., Cabibbo, N., Candelier, R., Cavagna, A., Cisbani, E., Giardina, I., Lecomte, V., Orlandi, A., Parisi, G., Procaccini, A., Viale, M., Zdravkovic, V.: Interaction ruling animal collective behavior depends on topological rather than metric distance: Evidence from a field study. Proc. Natl. Acad. Sci. USA **105**(4), 1232–1237 (2008)
11. Bellomo, N.: Modeling complex living systems – A kinetic theory and stochastic game approach. Modeling and Simulation in Science, Engineering and Technology. Birkhäuser, Boston (2008)

12. Bellomo, N., Carbonaro, B.: Toward a mathematical theory of living systems focusing on developmental biology and evolution: A review and perspectives. Phys. Life Rev. **8**(1), 1–18 (2011)
13. Bellomo, N., Herrero, M.A., Tosin, A.: On the dynamics of social conflicts: Looking for the Black Swan. Kinet. Relat. Models **6**(3), 459–479 (2013). Open Access http://dx.doi.org/10.3934/krm.2013.6.459
14. Bellomo, N., Knopoff, D., Soler, J.: On the difficult interplay between life, "complexity", and mathematical sciences. Math. Models Methods Appl. Sci. **23**(10), 1861–1913 (2013)
15. Bellomo, N., Soler, J.: On the mathematical theory of the dynamics of swarms viewed as complex systems. Math. Models Methods Appl. Sci. **22**(suppl. 1), 1140,006 (29 pages) (2012)
16. Bertotti, M.L., Delitala, M.: Conservation laws and asymptotic behavior of a model of social dynamics. Nonlinear Anal. Real World Appl. **9**(1), 183–196 (2008)
17. Bertotti, M.L., Delitala, M.: On the existence of limit cycles in opinion formation processes under time periodic influence of persuaders. Math. Models Methods Appl. Sci. **18**(6), 913–934 (2008)
18. Bertotti, M.L., Delitala, M.: Cluster formation in opinion dynamics: a qualitative analysis. Z. Angew. Math. Phys. **61**(4), 583–602 (2010)
19. Bertotti, M.L., Modanese, G.: From microscopic taxation and redistribution models to macroscopic income distributions. Phys. A **390**(21–22), 3782–3793 (2011)
20. Bisi, M., Spiga, G., Toscani, G.: Kinetic models of conservative economies with wealth redistribution. Commun. Math. Sci. **7**(4), 901–916 (2009)
21. Comincioli, V., Della Croce, L., Toscani, G.: A Boltzmann-like equation for choice formation. Kinet. Relat. Models **2**(1), 135–149 (2009)
22. Cristiani, E., Piccoli, B., Tosin, A.: Modeling self-organization in pedestrians and animal groups from macroscopic and microscopic viewpoints. In: G. Naldi, L. Pareschi, G. Toscani (eds.) Mathematical Modeling of Collective Behavior in Socio-Economic and Life Sciences, Modeling and Simulation in Science, Engineering and Technology, pp. 337–364. Birkhäuser, Boston (2010)
23. Düring, B., Markowich, P., Pietschmann, J.F., Wolfram, M.T.: Boltzmann and Fokker-Planck equations modelling opinion formation in the presence of strong leaders. Proc. R. Soc. Lond. Ser. A Math. Phys. Eng. Sci. **465**(2112), 3687–3708 (2009)
24. Goyal, S., Vega-Redondo, F.: Network formation and social coordination. Game Econ. Behav. **50**(2), 178–207 (2005)
25. Helbing, D.: Stochastic and Boltzmann-like models for behavioral changes, and their relation to game theory. Phys. A **193**(2), 241–258 (1993)
26. Helbing, D.: Social Self-Organization. Springer-Verlag, Berlin Heidelberg (2012)
27. Herbert, S.: Bounded rationality and organizational learning. Organ. Sci. **2**(1), 125–134 (1991)
28. Maldarella, D., Pareschi, L.: Kinetic models for socio-economic dynamics of speculative markets. Phys. A **391**(3), 715–730 (2012)
29. Naldi, G., Pareschi, L., Toscani, G. (eds.): Mathematical Modeling of Collective Behavior in Socio-Economic and Life Sciences. Modeling and Simulation in Science, Engineering and Technology. Birkhäuser, Boston (2010)
30. Rubinstein, A.: Modeling Bounded Rationality, Zeuthen Lecture Book, vol. 1. MIT Press, Cambridge, MA (1998)
31. Taleb, N.N.: The Black Swan: The Impact of the Highly Improbable. Random House, New York City (2007)
32. Toscani, G.: Kinetic models of opinion formation. Commun. Math. Sci. **4**(3), 481–496 (2006)

Using Mathematical Modelling as a Virtual Microscope to Support Biomedical Research

Chiara Giverso and Luigi Preziosi

Abstract This chapter will explain what kind of support mathematics can give to biology and medicine. In order to explain the concepts in practice cell migration is used as a specific example. This phenomenon is of great biomedical interest because it is a fundamental phenomenon both in physiological (e.g. wound healing, immune response) and pathological processes (e.g. chronic inflammation, detachment of metastasis and related tissue invasion). Also a key feature of any artificial system aimed at mimicking biological structures is to allow and enhance cell migration on or inside it. At the same time anti-cancer treatment can become more efficient blocking cell's capability to migrate towards distant sites and invade different organs.

1 Mathematical models in biomedicine

In [3] J.E. Cohen foresaw for this century a brilliant future for the interactions between Mathematics and Medicine stating that Mathematics can be Biology's next microscope representing for Biology what it meant for Physics in the last century. Already at the very beginning of the century the decoding of the human genome and the start of what can be called the post-genomic era represented a test bench for this conjecture. In fact, biologists started being swamped by an overwhelming quantity of data to handle and organise. More importantly they have now to unravel the underlying message and understand the link between gene expression, protein production or inhibition and cell behaviour. In order to do that in an efficient way

C. Giverso
Dipartimento di Scienze Matematiche, Politecnico di Torino, Corso Duca degli Abruzzi 24, 10129 Torino, Italy
e-mail: chiara@giverso.com

L. Preziosi (✉)
Dipartimento di Scienze Matematiche, Politecnico di Torino, Corso Duca degli Abruzzi 24, 10129 Torino, Italy
e-mail: luigi.preziosi@polito.it

A. Celletti, U. Locatelli, T. Ruggeri and E. Strickland (eds.): *Mathematical Models and Methods for Planet Earth.* Springer INdAM Series 6, DOI 10.1007/978-3-319-02657-2_5,

they soon realised that they had to rely on multidisciplinary collaborations to use non standard mathematical and statistical tools.

This represented a "chance" to respond to other classical biomedical needs characterizing bio-medical research. The first one consists in understanding starting from some diagnostic images and data collected during a certain observation period: (i) how the specific pathological condition developed and (ii) foresee the possible developments of the pathology in the possible different scenarios. Mathematically speaking the former is an inverse problem and the latter is a direct problem. A further need is to optimise the type of intervention and to identify the best therapeutic protocol on the specific patient. This is an optimal control problem that need to be based on reliable mathematical models.

It is then clear why there is a need to develop mathematical models that might lead to reliable numerical simulations and how these might help optimising the protocols avoiding lengthy, costy, and improductive trial-and-errors procedures. In this way Mathematics may act as a decision support system.

One has to keep in mind that bio-medical research already develops under the concept of modelling, that in this case is represented by biological experiment either in vitro or in vivo (for instance, the use of immuno-depressed mice instead of humans). In fact, going from clinics to labs the general procedure consist in trying to simplify the phenomenon focusing on those aspects that are considered fundamental for the understanding of the process, hoping that the crucial aspects are retained in the biological model. This procedure is not far from that used to develop a mathematical model starting from the phenomenological observation.

In fact, the high complessity characterizing all living systems and the uncountable relations existing among their many components make it impossible to describe biological systems in full. They require to focus on specific sub-phenomena and the introduction of drastically simplified models, both biological and mathematical.

In biomedical research, the phenomenological observation of a pathology may suggest some experiments to be performed in vivo, ex vivi, or in vitro. The first are the closest to the real observation on the patient but are also heavy from the ethical viewpoint. The last are the most flexible, controllable, reproducible from the experimental viewpoint, and relatively harmless.

Simulating the phenomena of interest can also help optimising the experiments, helping in writing down a priority list that consists in identifying in advance the most promising experiments and therapeutic factors and at the other extreme the less influencial strategies.

The crucial step in the modelling procedure sketched in Figure 1, is to pass from a good model that is able to simulate known biological facts to the simulation of unknown scenarios, of virtual drugs, not discovered yet, or of situations that have not been tested yet, or can not even be tested. The advantage of supporting biomedical research with mathematical models in this respect relies on the fact that in principle spanning over the entire range of parameters can be done much faster than the time required to prepare the experiments and to wait for the results.

In the rest of this chapter we will specify the support that mathematics can give to medicine focusing on some studies done on cell migration.

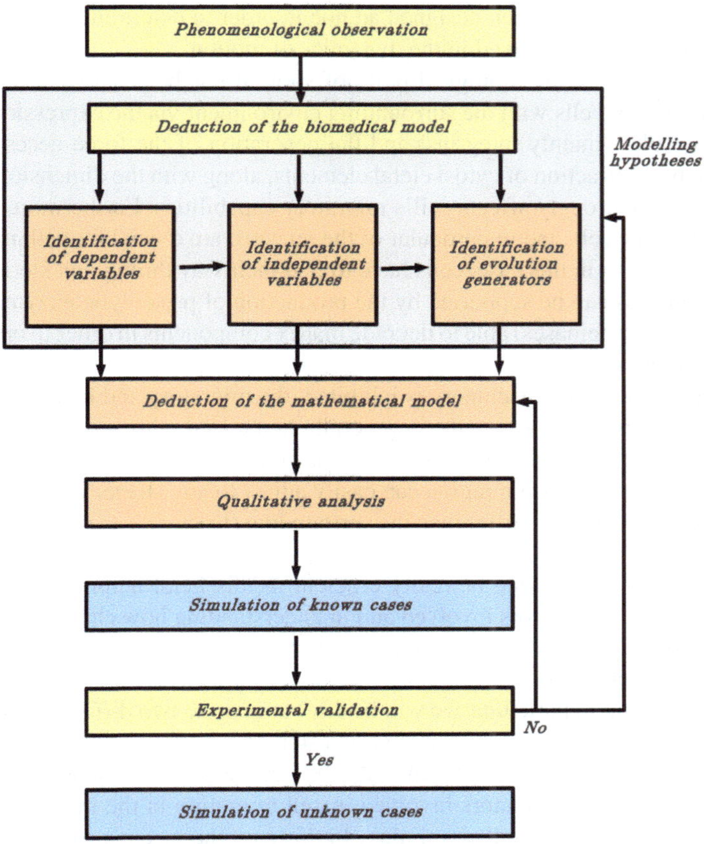

Fig. 1 The mathematical modelling cycle in biomathematics. The different colours refer to experimental actions (yellow), modelling (orange), numerical (blue)

2 Application to cell migration and invasion

As an example we will base our reasoning on cell migration because this is a fundamental phenomenon both in physiological (e.g. wound healing, immune response) and pathological processes (e.g. chronic inflammation, detachment of metastasis and related tissue invasion).

Also a key feature of any artificial system aimed at mimicking biological structures consists in allowing and enhancing cell migration on or inside it. At the same time anti-cancer treatment can become more efficient blocking cell's capability to migrate towards distant sites and invade different organs.

The importance of cell migration on substrates and through biological barriers and the determination of the different factors involved in such a complex process is a well established subject in the biological community. Indeed, in recent years, experimental settings have been designed ad hoc in order to determine cell properties and functions that are involved in the dynamics of motion.

In particular, from the biological point of view, it has been shown that the dynamic adhesion of cells with the surrounding environment via the expression of adhesive molecules (mainly integrins) and the generation of the force necessary for propulsion by contraction of cytoskeletal elements, along with the dimensionality of the environment strongly affects cell's migratory capabilities. Furthermore, the deformability of the cell, and in particular of the nucleus, are crucial for cell migration in 3D structures, as in this case, cells have to find their way throughout steric obstacles. This process can be supported by the production of proteolytic enzymes (e.g., matrix metallo-proteinases) able to degrade matrix components in order to open gaps in the environment.

When the proteolytic machinery is inhibited, the migratory and invasive process in three-dimensional environments is generally associated with significant cell and nucleus deformation while passing through constrictions in the extracellular matrix. In principle, the cytoplasmic region can easily adjust to any shape, whereas the nucleus, which is 5-10 times stiffer than the surrounding structure, can resist to changes in shape.

Mathematical models can be really efficient in this field, helping biologists in studying the different factors involved and in understanding how changes in cell capability to adhere to the substrate, contract, deform and secrete proteolytic enzymes can affect the overall process.

In the following in particular we will briefly summarize two different mathematical models related to cell migration and invasion of the surrounding environment. In the first application we will present a discrete model aimed at understanding the interplay of the different factors involved in cell migration in the invasive process of cancer cells into the mesothelial lining. In this case the mechanical properties of the nucleus are not considered, because the process is associated with intense secretion of matrix metallo-proteinases and retraction of the cell composing the biological barrier.

On the other hand, in the second application we will focus on the process of a cell, with inhibited ability to secrete matrix metallo-proteinases into cylindrical channels composed by extracellular matrix. In this case, the environment is supposed rigid and thus nucleus deformability is a limiting factor in the process of migration.

2.1 A cellular potts model for ovary cancer invasion

A crucial process in the growth of a cancer, which makes it so difficult to be treated, is the capability of cancer cells to reach distant organs in the body and form metastases. For instance, ovarian cancer cells are able to form widespread intraperitoneal dissemination throughout the abdomen and the pelvis. The process of metastasization involves exfoliation of tumor cells as single cells or aggregates from the primary tumor located into the abdominal cavity, and the successive implantation on and invasion through the mesothelial lining of the peritoneum. Mathematical models can help understand the process of transmigration of ovarian cancer cells through the mesothelial layer and direct biological experiments. In particular, to reproduce the typical in vitro experiments used to study this process, both for single cells and for cell aggregates or spheroids the process can be intuitively presented using a Cellular Potts Model (CPM) [4]. CPMs can well capture mechanisms of cellular adhesion, motion and the degradation of the extracellular matrix, interfacing with reaction-diffusion models describing the diffusion and uptake of chemotactic factors and for the release of matrix metalloproteinases from tumor cells.

The CPM is a grid-based stochastic approach in which each cell is represented by a connected set of grid-sites [6, 8, 12, 13]. The behaviour of the cells and their interactions, both mutual and with the local microenvironment, is then described in energetic terms and constraints.

Referring to [17, 18] for more details, the simulation domains is a d-dimensional regular lattices $\Omega \subset \Re^d$, where $d = 2,3$ according to the specific application. Each lattice site $\mathbf{x} \in \Omega \subset \Re^d$ is labeled by an integer number, $\sigma(\mathbf{x})$. As classically adopted in CPM applications, a neighbour of $\mathbf{x}$ is identified by $\mathbf{x}'$, while its overall neighbourhood by $\Omega_{\mathbf{x}}'$, i.e. $\Omega_{\mathbf{x}}' = \{\mathbf{x}' \in \Omega : \mathbf{x}' \text{ is a neighbour of } \mathbf{x}\}$. Subdomains of contiguous sites with identical σ form discrete objects Σ_σ (i.e., $\Sigma_{\overline{\sigma}} = \{\mathbf{x} \in \Omega : \sigma(\mathbf{x}) = \overline{\sigma}\}$), which are characterized by an object type, $\tau(\Sigma_\sigma)$.

The keypoint is then to identify a hamiltonian, whose minimization will drive the evolution of the cellular system. It may be composed of many terms, some essential, others optional, that sum up to form the full hamiltonian.

An essential term governs the geometrical attributes of simulated objects such as cell size and stiffness and can be written as:

$$H_{shape}(t) = H_{volume}(t) + H_{surface}(t) = \\ = \sum_{\Sigma_\sigma} \left[\kappa_{\Sigma_\sigma}(t) \left(\frac{v_{\Sigma_\sigma}(t) - V_{\tau(\Sigma_\sigma)}}{v_{\Sigma_\sigma}(t)} \right)^2 + \nu_{\Sigma_\sigma}(t) \left(\frac{s_{\Sigma_\sigma}(t) - S_{\tau(\Sigma_\sigma)}}{s_{\Sigma_\sigma}(t)} \right)^2 \right]. \quad (1)$$

The form of the hamiltonian depends on the actual volume and surface of the object, $v_{\Sigma_\sigma}(t)$ and $s_{\Sigma_\sigma}(t)$ (which reduce, respectively, to its surface and perimeter in two dimensions), as well as on the same quantities in the relaxed state, $V_{\tau(\Sigma_\sigma)}$ and $S_{\tau(\Sigma_\sigma)}$, corresponding to its initial measures. The terms $\kappa_{\Sigma_\sigma}(t)$ and $\nu_{\Sigma_\sigma}(t) \in \Re^+$ represent mechanical moduli in units of energy: in particular, $\kappa_{\Sigma_\sigma}(t)$ refer to volume changes, while $\nu_{\Sigma_\sigma}(t)$ relates to the degree of deformability/elasticity of the related object, i.e. the ease with which it is able to remodel. Indeed, assuming that cells do not significantly grow during migration, the fluctuations of their volumes are kept negligible with high constant values $\kappa_{\Sigma_\sigma} = \kappa \gg 1$.

Another essential term describes the adhesive interaction between cells or between a cell and a matrix component that can be written as:

$$H_{adhesion}(t) = \sum_{\substack{\mathbf{x} \in \Omega, \mathbf{x}' \in \Omega_{\mathbf{x}}' \\ \Sigma_{\sigma(\mathbf{x})} \neq \Sigma_{\sigma(\mathbf{x}')}}} J_{\tau(\Sigma_{\sigma(\mathbf{x})}), \tau(\Sigma_{\sigma(\mathbf{x}')})}. \quad (2)$$

The Js are binding energies per unit area, which are obviously symmetric and measure of the affinity between cell surface adhesion complexes on the membrane of the two cells or with extracellular ligands.

Being more specific, usually cells reside in an extracellular matrix, which is differentiated in a medium-like state, $\tau = M$, and a fibrous-like state, $\tau = F$. The medium-like state reproduces the mixture of soluble components, which, together with the water solvent, compose the so called interstitial fluid. The fibrous-like state represents instead the network of insoluble macromolecules, such as collagens, laminin and elastin, that associate into fibers. Coming to the binding energies, $J_{C,F} < J_{C,M}$ since, as widely demonstrated in literature, most cell lines in standard conditions adhere more strongly with the fibrous part of the extracellular matrix than with its soluble component. $J_{C,C}$ is instead kept high to avoid cell-cell adhesive interactions that may affect their movement. By setting constant and homogeneous values for the bond energies Js, we here assume a uniform distribution of adhesion molecules on cell surfaces and of ligands in the external environment, without any change during the observation time.

Other optional terms in the hamiltonian might be introduced to model chemotaxis, i.e., the motion towards higher concentrations of specific chemical factors, or persistence, i.e., the tendency of polarized cells to keep their direction of motion.

The CPM can be easily implemented using CompuCell3D[1] package [6, 9], an open source modeling environment and pde solver, that allows users to easily define their model. One of the advantages of CPM is to obtain a morphological description of the process, that can be easily compared to biological experiments. Furthermore the model allows to perform a systematic study of the different factors involved, in the biological phenomenon. In particular, the simulation obtained in [4] are able to reproduce surprisingly well the morphology observed during biological experiments performed by N. Lo Buono (Laboratory of Immunogenetics, Department of Genetics, Biology and Biochemistry, Torino), letting understand the importance of the different families of adhesion molecules (i.e., cadherins and integrins) in the overall ovarian cancer transmigration process.

In this case, the mathematical simulations can help in the definition of the principal factors to be studied through biological experiments. Indeed, through the mathematical model it is easy to study the relative importance of the different factors involved in the process, suggesting to biologists the direction of future works (see, for instance, [10, 13]). The numerical simulations shows that changes in the expression of selected adhesive molecules, along with the release of tumor matrix metalloproteinases and the consequent degradation of extracellular matrix components are the key players of the overall process [4]. In particular, it is demonstrated that a crucial role in the transmigration of single cells is played by matrix metalloproteinases, which are responsible of the extracellular matrix degradation, while the adhesion affinity between tumor cells and mesothelial cells is crucial to determine the inclusion of the tumor cell in the mesothelial layer. In addition to the previous phenomena in the case of the spheroid invasion, the relative adhesion affinity between the tumor

[1] http://www.compucell3d.org

cells and between them and the mesothelial layer determines whether the invasion is monofocal or multifocal. In any case, it is found [4] that mesothelial invasion by single cells is more conservative, while the invasion of tumor spheroids causes the disruption of the mesothelium. Indeed isolated cells can rapidly change their shape to cross the mesothelium through gaps opened by the activity of the MMPs, but, once the transmigration is completed, the mesothelial junctions are recovered and the continuity of the layer is restored (see the left column in Fig. 2). On the other hand, cancer spheroids, completely overtake larger areas of the mesothelium, forming different foci of invasion and inducing apoptosis of the detached part of the mesothelium (see the right column in Fig. 2).

The CPM allows to easily study the influence of different parameters in the overall process, in particular, the influence of matrix metalloproteinases secretion ($\overline{\vartheta}_{MMP}$), of the contact energy between mesothelial cells (J_{τ_m,τ_m}), and of the contact energy between cancer and mesothelial cells (J_{τ_c,τ_m}). Specifically, referring to the top panel on the left of Figure 2:

- For low metalloproteinases activity (low $\overline{\vartheta}_{MMP}$) and for high affinity of mesothelial cells (low J_{τ_m,τ_m}) the cell can not pass thorugh the mesothelial lining and invade the tissue below.
- For high affinity of cancer cells with mesothelial cells (low J_{τ_c,τ_m}) the cell gets trapped in the mesothelial layer and will subsequently replace the physiological tissue with a pathological one.

The simulations give an indication of the morphology acquired by cells during the process (qualitative behaviours) and of the time (in Monte Carlo Steps, MCS) required to accomplish the process of transmigration. The success of cellular Potts models and of other individual cell-based models in the biological community is related to the fact that biologists and medical doctors see on the monitor something similar to what they see in the microscope.

Of course, in order to obtain more details on the microscopic biological processes involved, sub-cellular pathways governing the expression and the activity of adhesion molecules and the secretion of the matrix metalloproteinases have to be introduced in the model. Introducing the microscopic level of description and linking it to the mesoscopic or the macroscopic level is a fundamental step from the biomedical point of view. In fact, this is the level on which most attention is paid for drug discovery and for the identification of potential therapies. Some efforts in introducing subcellular pathways in the discrete cellular model have been done as described in [18] and references therein.

2.2 The role of nuclear mechanical properties

In [4] no distinction between the different compartments composing the cell body (i.e. the nucleus and the cytoplasm at least) have been made. This particular improvement is essential when the channel size has the same dimension as the nucleus

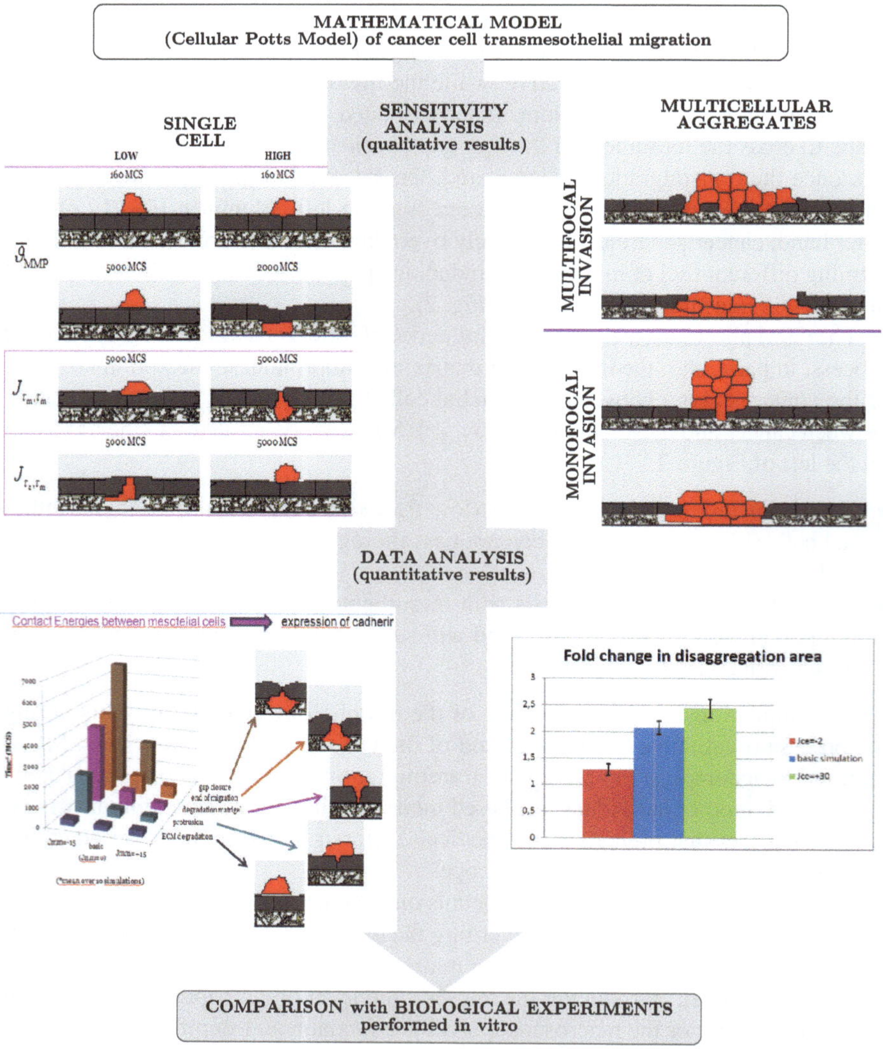

Fig. 2 From the formulation of the mathematical model to the validation of the model by comparing the qualitative and quantitative results obtained with the CPM for a single cell transmesothelial migration

that represents the most rigid element of the cell, because the cell can get stuck in the network. Still using cellular Potts models this was done in [19,20] analyzing the role of rigidity of both the extracellular matrix and the nucleus on cell migration.

The main outcome is represented in Figure 3 and consists in evaluating the cell speed as a function of the mechanical and geometrical characteristics of the cells and of the fibrous environment they live in. In particular, the simulations put in evidence the existence of an optimal ratio of the dimension of the nucleus versus the pore size

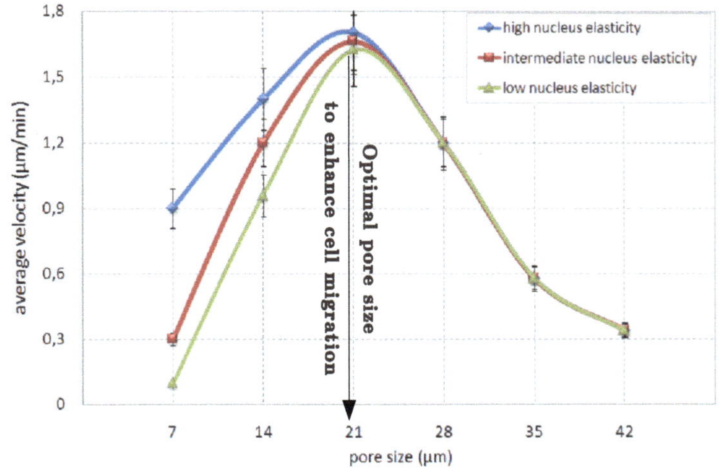

Fig. 3 Dependence of cell velocity from the pore size of the extracellular matrix

of the network to achieve faster migration. This result can be used to improve the manufacturing of scaffolds used for wound healing that need to be populated fast by fibroblasts and keratinocytes that operate to close the wound.

For what concerns continuum models, a first attempt to include nucleus mechanical properties into the description of cell migration inside 3D substrates has been done in [5], addressing the problem of cell entry into cylindrical extracellular matrix structures.

A proper representation of the deformation experienced by the nucleus, treated as an incompressible neo-Hookean elastic solid, is obtained in order to compute the total energy required to deform the nucleus from the initial to the final deformed shape. This energy is then compared with the total work done by active forces, generated after the adhesion of the cell to the surrounding extracellular matrix (integrin-dependent migration). The adhesion of the cell with the substrate is fundamental in order to activate the actomyosin contraction necessary for nucleus deformation and cell movement along the track [22]. The traction force is related to the adhesion area to its location. In fact, it is known that adhesion sites are particularly active in the frontal part of the cell. Very recently it was possible to quantify this qualitative observation by solving an inverse problem [1, 21], that, given the deformation of the network on which (or in which) the cell move, was able to compute the force that the cell need to exert on the focal adhesions distributed on the membrane to cause the measured deformation. This is another example in which a mathematical tool has been used to get insight into the biological properties, specifically to transform indirect experimental measurements on substratum deformation into theoretical evaluation of the traction forces.

After postulating the mathematical model for active forces required to accomplish the process of migration inside the channel, some energy-based criteria are identified. The energetic criteria proposed in [5] leads to the identification of some

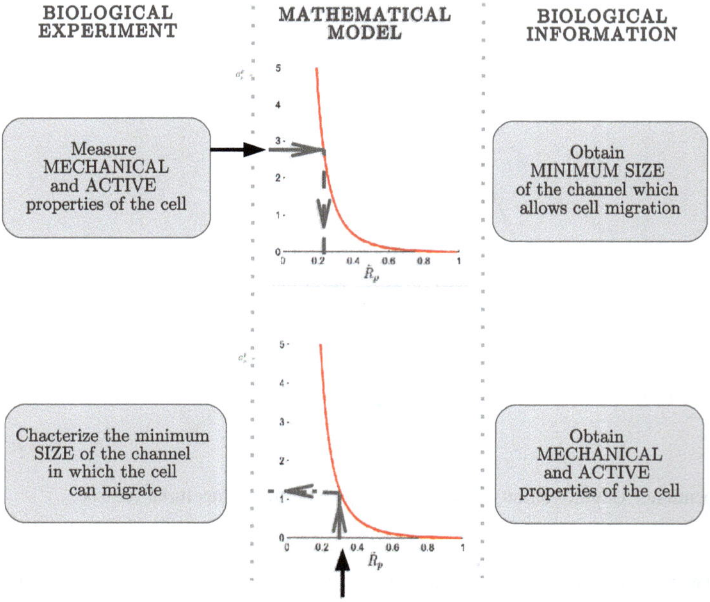

Fig. 4 Importance of deriving a mathematical model able to consider nuclear mechanical properties in the process of migration. $\tilde{R}_p = R_p/R_n$ is the ratio of the microchannel size versus the nucleus diameter. A mathematical model able to capture the influence of nuclear mechanical properties on the process of cell migration can be very useful in order to obtain the minimum size of the channel which allow cell migration, once that mechanical and active properties of the cell are known, or in order to derive the ratio between active and mechanical properties of a cell, starting from biological experiments of cell migration inside channel of different size

characteristic parameters, which take into account the mechanical properties of cell nucleus, the adhesive characteristics of the cell membrane, the active force generated through cytoskeleton contraction and the aspect ratio between the channel radius and the radius of the undeformed nucleus. In particular, a key dimensionless parameter is $G_\mu^F = \rho_b \alpha_{ECM} F_b^M/\mu$ where ρ_b is the density of bonds, α_{ECM} is the extracellular matrix ratio, F_b^M is the maximum force exerted on the nucleus, μ is the shear elastic modulus of the nucleus. Summarizing G_μ^F represents the ratio between cell active/adhesive properties (i.e., the capability of the cell to form adhesive sites and contract) and the mechanical properties of the nucleus.

The analytical and semi-analytical relations proposed in [5] provide the relations between active and mechanical properties that should be satisfied in order to have cells entering a channel of given radius. Therefore, knowing the adhesive, mechanical and contractile properties of the cell, it is possible to derive the minimum channel size and, conversely, observing experimentally the capability of a cell to enter cylindrical channels of different dimensions, it is possible to characterize the interplay between mechanical and active properties (see Fig. 4).

The results predict that cells are able to enter extracellular matrix-networks only for pore radii bigger than a critical value, that can be determined a priori, depending on the stiffness of their nucleus and their capabilities of expressing adhesion molecules in order to bind to the extracellular matrix. The mathematical findings are in good agreement with the biological observation that a rigid cell body would nullify any attempt of the cell to squeeze through channels and network gaps narrower than the nucleus dimension, as observed in [15, 22].

Therefore the mathematical model presented in [5] suggests that mechanical properties of the nucleus are fundamental in the process of migration, and that, blocking the capability of the nucleus to deform can potentially inhibit cell capability to migrate in the surrounding environment.

From the biomedical point of view, the analytical relations obtained in [5] and the results in [19, 20] could be applied to the design of synthetic scaffolds, with optimal values of pore size and fibre density, that may accelerate cell transport and in-growth, critical for regenerative treatments (see Fig. 4).

3 Multiscale modelling

It is clear that cell and tissue behaviour depends on the interactions they have with the environment. The latter depends in turn on several chemical cues, e.g., genetic information, gene expression, activation of particular signaling pathways. Consequently, even if one is interested only in describing a biological phenomenon from the macroscopic point of view, it becomes natural and fundamental to include into the models at the tissue or cellular scale also the processes occurring at the sub-cellular scale, e.g., the activation of specific protein cascades.

For this reason, some models developed to describe biomedical processes are nowadays paying more and more attention to the chemical phenomena inside the cell, nesting in the macroscopic (or mesoscopic) representation one or more modules accounting for processes at the microscopic scale. Thanks to the most recent discoveries in the field of genomics, proteomics, and system biology, these descriptions are expanding considerably, and, in our opinion, their use will increase more and more.

In this respect, models at the cellular scale, such as the cellular Potts models briefly described above, seem more flexible and suited to include sub-cellular mechanisms.

However, individual cell-based models and models based on partial differential equations are not in principle in alternative as each one has its advantages and disadvantages. For instance, the latter are more suited and computationally cheaper to describe the behaviour of tissues from the macroscopic point of view, but, as discussed above, it is more difficult to account for sub-cellular mechanisms such as signal transduction, expression or internalisation of receptors, and so on. Conversely, individual cell-based models can focus more closely on the cellular level, but they might become computationally expensive when using a large number of cells, and inefficient in providing a general outlook on the system as a whole.

Luckily, it is often not necessary to keep the same level of detail throughout the tissue. The idea behind what can be called *interfacing hybrid models* is to use the modelling tools like a microscope, focussing on the cellular or sub-cellular level where and when needed and blurring when such a detailed description is not needed. The idea is to split the domain in several time-dependent sub-domains, and in using in each piece of this moving puzzle a different modelling framework, e.g. cellular Potts models and continuous models, so that it is possible to take advantage of the positive aspects of both frameworks.

The first step in this process is to compare, for the same biological phenomenon, the results obtained via different models. This way, it is possible to support the up-scaling process and get an idea of the relationship between those parameters that cannot be easily linked.

The next step would be to get the mechanical behaviour of non-growing tissues described using individual cell-based models from standard virtual mechanical tests, e.g., stress relaxation, creep and indentation tests, steady shear, and cyclic loading and deformations. In our opinion, this might be done by building *in silico* experiments, like those performed in rheology, for instance using cone-and-plate rheometers. Ensembles of cells with some given microscopic characteristics (e.g., cell-cell adhesion or cell compressibility) could then be put in these virtual experiments and tested. One could get as output the macroscopic parameters to be used in the constitutive equation of the multiphase model, that hopefully guarantee the preservation of the mechanical properties when switching back and forth from tissues to macroscopic models.

All this must be supported by suitable mathematical procedures that allow models to cross-talk and is a big challenge for the future. In the meantime, other hybrid models have been developed. The most common one uses a discrete approach (e.g., discretization of partial differential equations, cellular automata, individual cell-based models, cellular Potts models) for some constituents and a continuous approach (e.g., diffusion equations or more general partial differential equations) for other constituents such as proteins and molecules in general. Typically, cells are described using spatially discrete variables, while molecules and extracellular matrix are described using spatially continuous variables. A brief review of different hybrid models can be found in [2, 14] who also address what are the biggest problems to face in the near future to have handy models that are closer and closer to the real biological phenomenology and can be a real support for the development of medicine.

References

1. Ambrosi, D.: Cellular traction as an inverse problem. SIAM J. Appl. Math. **66**, 2049–2060 (2006)

2. Anderson, A.R.A., Chaplain, M.A.J., Rejniak, K.A.: Single-cell-based models in biology and medicine. Mathematics and Biosciences in Interaction. Birkhäuser-Verlag, Basel (2007)
3. Cohen, J.E.: Mathematics is biology's next microscope, only better; Biology is mathematics' next physics, only better. PLoS Biology **2**, e439 (2004)
4. Giverso, C., Scianna, M., Preziosi, L., Lo Buono, N., Funaro, A.: Individual cell-based model for in-vitro mesothelial invasion of ovarian cancer. Math. Model. Nat. Phenom. **5**, 203–223 (2010)
5. Giverso, C., Grillo, A., Preziosi, L.: Influence of nucleus deformability on cell movement into cylindrical structures, Biomech. Model. Mechanobiol. (2013). doi 10.1007/s10237-013-0510-3
6. Glazier, J.A., Graner, F.: Simulation of the differential adhesion driven rearrangement of biological cells. Physical. Rev. E **47**, 2128–2154 (1993)
7. Glazier, J. A., Balter, A., Poplawski, N.J.: Magnetization to morphogenesis: a brief history of the Glazier–Graner–Hogeweg model. In: Anderson, A.R.A., Chaplain, M.A.J., Rejniak, K.A. (eds.) Single-Cell-Based Models in Biology and Medicine. Mathematics and Biosciences in Interactions, pp. 79–106. Birkhäuser-Verlag, Basel (2007)
8. Graner, F., Glazier, J.A.: Simulation of biological cell sorting using a two-dimensional extended Potts model. Phys. Rev. Letters **69**, 2013–2017 (1992)
9. Hentschel, H.G.E., Glimm, T., Glazier, J.A., Newman S.A.: Dynamical mechanisms for skeletal pattern formation in the vertebrate limb. Proc. R. Soc. Lond. B **271**, 1713–1722 (2004)
10. Lo Buono, N., Parrotta, R., Morone, S., Bovino, R., Nacci, G., Ortolan, E., Horenstein, A.L., Inzhutova, A., Ferrero, E., Funaro, A.: The CD157-integrin partnership controls transendothelial migration and adhesion of human monocytes, J. Biol. Chem. **286**, 18681–18691 (2011)
11. Marée, A.F.M., Grieneisen, V.A., Hogeweg, P.: The Cellular Potts Model and biophysical properties of cells, tissues and morphogenesis. In: Anderson, A.R.A., Chaplain, M.A.J., Rejniak, K.A. (eds.) Single-Cell-Based Models in Biology and Medicine. Mathematics and Biosciences in Interactions, pp. 107–136. Birkhäuser-Verlag, Basel (2007)
12. Merks, R.M.H., Glazier, J.A.: A cell-centered approach to developmental biology. Physica. A. **352**, 113–130 (2005)
13. Morone, S., Lo Buono, N., Parrotta, R., Giacomino, A., Nacci, G., Brusco, A., Larionov, A., Ostano, P., Mello-Grand, M., Chiorino, G., Ortolan, E., Funaro, A.: Overexpression of CD157 contributes to epithelial ovarian cancer progression by promoting mesenchymal differentiation. PLoS ONE (2012). doi:10.1371/journal.pone.0043649
14. Preziosi, L., Tosin, A.: Multiphase and multiscale trends in cancer modelling, Math. Model. Nat. Phenom. **4**, 1–11 (2009).
15. Rolli, C.G., Seufferlein, T., Kemkemer, R., Spatz, J.P., Impact of tumor cell cytoskeleton organization on invasiveness and migration: A microchannel-based approach. PLos ONE **5**, e8726 (2010)
16. Savill, N.J., Hogeweg, P.: Modelling morphogenesis: from single cells to crawling slugs, J. Theor. Biol. **184**, 118–124 (1997)
17. Scianna, M, Preziosi, L.: Multiscale developments of cellular Potts models. Multiscale Model. Simul. **10**, 342–382 (2012)
18. Scianna, M, Preziosi, L.: Cellular Potts Models: Multiscale Developments and Biological Applications, Chapman & Hall/CRC Press (2013)
19. Scianna, M., Preziosi, L.: Modelling the influence of nucleus elasticity on cell invasion in fiber networks and microchannels, J. Theor. Biol. **317**, 394–406 (2013)
20. Scianna, M., Preziosi, L., Wolf, K.: A cellular potts model: simulating cell-and extracellular matrix-derived determinants for cell migration on and in matrix environments, Math. Biosci. Eng. **10**, 235–261 (2013)
21. Vitale, G., Preziosi, L., Ambrosi, D.: A numerical method for the inverse problem of cell traction in 3D. Inv. Prob. **28**, 095013 (2012)
22. Wolf, K., Wu, Y.I., Liu, Y., Geiger, J., Tam, E., Overall, C., Stack, M.S., Friedl, P.: Multistep pericellular proteolysis controls the transition from individual to collective cancer cell invasion. Nat. Cell. Biol. **9**, 893–904 (2007)

Ferromagnetic Models for Cooperative Behavior: Revisiting *Universality* in Complex Phenomena

Elena Agliari, Adriano Barra, Andrea Galluzzi, Andrea Pizzoferrato and Daniele Tantari

Abstract Ferromagnetic models are *harmonic oscillators* in statistical mechanics. Beyond their original scope in tackling phase transition and symmetry breaking in theoretical physics, they are nowadays experiencing a renewal applicative interest as they capture the main features of disparate complex phenomena, whose quantitative investigation in the past were forbidden due to data lacking. After a streamlined introduction to these models, suitably embedded on random graphs, aim of the present paper is to show their importance in a plethora of widespread research fields, so to highlight the unifying framework reached by using statistical mechanics as a tool for their investigation. Specifically we will deal with examples stemmed from sociology, chemistry, cybernetics (electronics) and biology (immunology).

E. Agliari
Dipartimento di Fisica, Università di Parma, Viale Usberti 7A, I-43100 Parma, Italy
e-mail: elena.agliari@fis.unipr.it

A. Barra (✉)
Dipartimento di Fisica, Sapienza Università di Roma, P.le Aldo Moro 2, I-00185 Roma, Italy
e-mail: adriano.barra@roma1.infn.it

A. Galluzzi
Dipartimento di Matematica, Sapienza Università di Roma, P.le Aldo Moro 2, I-00185 Roma, Italy
e-mail: galluzzi@mat.uniroma1.it

A. Pizzoferrato
Dipartimento di Fisica, Sapienza Università di Roma, P.le Aldo Moro 2, I-00185 Roma, Italy
e-mail: pizzoferrato@fis.uniroma1.it

D. Tantari
Dipartimento di Matematica, Sapienza Università di Roma, P.le Aldo Moro 2, I-00185 Roma, Italy
e-mail: tantari@mat.uniroma1.it

A. Celletti, U. Locatelli, T. Ruggeri and E. Strickland (eds.): *Mathematical Models and Methods for Planet Earth.* Springer INdAM Series 6, DOI 10.1007/978-3-319-02657-2_6,

1 Introduction

The history of theoretical investigation in collective ferromagnetic behaviors is rooted back in the first decades of the twentieth century, when Lenz introduced a model and – in the winter of 1921 – asked to his student Ising to solve it as, for himself, such a research was too trivial. The Second World War, and in particular Nazi persecution, estranged Ising from Germany (and from scientific interchanges) up to late 1947, when, once back, somehow unexpectedly he discovered to be a famous physicist for his contribution in solving the Lenz model (which passed on as the Ising model [8, 15, 28]).

At that time statistical mechanics was developed as a theoretical tool for investigating the structure of matter (solid state physics, liquid and kinetic theories [7, 22]) and the first concept of "universality" [21] highlighted how different *physical* systems behave in a very similar way close to criticality. Other decades had to elapse before the scientific community started to realize that such "universal" behavior was far from being restricted to the physical scenario, and a mature understanding that "several element showing imitative interactions" may behave collectively as a ferromagnet – whatever the context – is nowadays achieved.

However, the boundaries of validity of the last assertion are still under investigation as our knowledge of ferromagnetism is growing and merging with "imitation" [20], "cooperation" [16], "amplification" [26], "syncronization" [1], etc. Moreover, disparate fields of sciences continually spring up: lapping (a part of) such boundaries is the focus of the present paper.

What we need, as a theoretical benchmark, is the description of an ensemble of dichotomic spins living on the nodes of random graphs. Hence, we first provide a streamlined introduction to the statistical mechanics of ferromagnetism and a minimal smattering regarding the underlying graph theory. Then, we turn to extrapolate such an imitative behavior from the real world, stemming examples from several fields of science as sociology, chemistry, cybernetics (electronics) and biology (immunology). Summarizing, we are going to show that:

- In sociology, focusing on a test-case among many [11, 20], namely the phenomenon of social integration of migrants inside a host community, we are going to analyze as a standard quantifier the amount of mixed marriages (where "mixed" means achieved by a native and a migrant): we will show that, once plotted against the percentage of migrants inside the host community, its behavior is identical to the one of observable typical of statistical mechanics (i.e. the magnetization versus the temperature), highlighting the key role – in imitative behavior – played by each agent belonging to the community. We will show how (and why) this phenomenon can be reabsorbed within the ferromagnetic phenomenology [12].
- In chemistry, in particular dealing with reaction kinetics as a concrete example, many polymers and proteins exhibit cooperativity, meaning that their ligands bind in a non-independent way: if, upon a ligand binding, the probability of further binding (by other ligands of the same protein/polymer) is enhanced, like in the paradigmatic case of hemoglobin [28], the cooperativity is said to be positive. As

we are going to show, such a cooperative behavior in chemical reactions can be perfectly described by the statistical mechanics of ferromagnetism [4, 19].

- In electronics, we find the hallmark of ferromagnetic behavior already in its fundamental bricks, namely in operational amplifiers [26]. As we will show, there is a one-to-one mapping between self-consistency in statistical mechanics and transfer function in electronics. In particular, when no amplification is present, such a transistor can be mapped into an ensemble of non-interacting spins, but, when the circuit is amplifying the input signal (hence the output is proportional to the input by a constant of proportionality larger than one) interaction among its constituents can be mapped into interaction among spins and, again, its behavior is perfectly described by means of the statistical mechanics of ferromagnetism [4].
- In immunology, B clones (namely the ensemble of identical B cells producing the same antibodies) can interact reciprocally by imitation (that immunologists call "elicitation"): if a clone undergoes expansion and antibody release, its nearest neighbor (in the idiotypic network, namely the random graph whose nodes are the B clones and whose links are their reciprocal strengths of interaction [10,25]) will also undergo clonal expansion and antibody release too. Again, such a behavior is remarkably captured by the statistical mechanics of ferromagnetism [6].

Before proceeding, we notice that the three-dimensional Ising model is still unsolved, and enormous efforts have been necessary, e.g. by Onsager and followers [15, 27], in order to solve the model at low dimensionality. However, for all our examples, and away from the physical world (where the power-laws of gravity and electromagnetic fields strongly require projection on two- and three-dimensional structures), we will deal with the so called "mean-field" approximation. The latter is completely solvable as it assumes spins interacting broadly on random graphs (e.g. Erdös-Rényi topologies [17]) instead of peer-to-peer physical interactions on lattices: while this feature constitutes an approximation in the pure-physical community, in all the branches of science we outline (as well as in several others), where interactions are not short-ranged, this is perfectly reasonable, at both theoretical and empirical levels. Indeed, the mean-field statistical mechanics, revealed itself as a powerful and unifying instrument to investigate the complexity of our world: our understanding of collective behaviors by interacting agents from this perspective is an extremely exciting research field, still at the beginning, and we believe statistical mechanics will become a stronger and stronger technology for this task in the near future.

2 Definition of the model and thermodynamics

Let us consider an ensemble of N agents (spins), whose state is represented by a dichotomic variable $\sigma_i = \pm 1$, with $i \in (1, \ldots, N)$; through the paper, spins will assume a different meaning according to the context. Agents interact with each other, if reciprocally connected, via a positive coupling J, hence we can write an Hamiltonian for the system as

$$H_N[\sigma|J] = -\frac{1}{N}\sum_{i<j} J_{ij}\sigma_i\sigma_j - h\sum_i \sigma_i, \tag{1}$$

where h is an external scalar field (magnetic in the physical literature) and the coupling J_{ij} is set equal to either 1 or 0 according to a given probability distribution. This choice automatically frames the model on an Erdös-Rényi graph [2]. By imposing $J_{ij} = 1$, when the link between i and j exists, we only lock the temperature scale without changing the physics of the problem. Of course $J_{ij} = 0$ simply means that the two corresponding nodes are not interacting.

The role of dilution, from a statistical mechanics perspective, at least at the level of the mean values of observable, is simply to reduce the averaged strength reciprocally felt by the spins, but does not alter[1] the physics [9, 18].

The thermodynamic of the model is carried by the free energy density $\lim_{N\to\infty} f_N(\beta) = \lim_{N\to\infty} F_N(\beta)/N$, which is related to the Hamiltonian via

$$e^{-\beta F_N(\beta,h)} = Z_N(\beta,h) = \sum_{\{\sigma\}} e^{\frac{\beta}{N}\sum_{i<j}\sigma_i\sigma_j + \beta h\sum_i \sigma_i}, \tag{2}$$

where $Z_N(\beta,h)$ is the partition function. For the sake of convenience we will not deal with $f_N(\beta,h)$ but with the thermodynamic pressure $A(\beta,h)$ defined via

$$A(\beta,h) = \lim_{N\to\infty} A_N = \lim_{N\to\infty} -\beta f_N(\beta,h) = \lim_{N\to\infty}\frac{1}{N}\ln Z_N(\beta,h). \tag{3}$$

A key role will be played by the order parameter, namely the magnetization m, that reads as

$$m_N = \frac{1}{N}\sum_{i=1}^{N}\sigma_i, \qquad \langle m_N\rangle = \frac{\sum_{\{\sigma\}} m_N e^{-\beta H_N[\sigma]}}{\sum_{\{\sigma\}} e^{-\beta H_N[\sigma]}}, \tag{4}$$

where in the last definition the brackets $\langle . \rangle$ denote the Boltzmann average.

Note that the order parameter, namely a single function of the tunable parameters that describes the "typical" behavior of the system, is nothing but the arithmetic average of all the single degrees of freedom the system may use to respect thermodynamics.

In order to solve for the free energy (strictly speaking for the pressure), namely to obtain an explicit functional expression of $A(\beta,h)$ in terms of the tunable parameters β and h and of the order parameter m, we are going to use Guerra's interpolation scheme [8, 18, 24]. The idea behind this approach is to interpolate between the original system and a system of independent spins interacting with an effective field able to simulate fictitiously the stimuli induced by the others. To this task we introduce the following interpolating Hamiltonian

$$H(t) = tH_{original} + (1-t)H_{one-body}, \tag{5}$$

[1] As far as the network remains over-percolated. If the percolation threshold is crossed, the system splits into independent subsystems and the analysis reduces to the sum of the analysis on each subsystem.

where $t \in [0,1]$ is the interpolation parameter, and the corresponding (time dependent) partition function $Z_N(t)$, pressure $A_N(t)$ and Boltzmann state $\langle m_N \rangle_t = Z_N^{-1}(t) \sum_{\{\sigma\}} m_N e^{-\beta H(t)}$. Choosing $H_{one-body}[\sigma] = -\bar{m} \sum_i \sigma_i$, once introduced a trial parameter $\bar{m}$ to be optimized at the end of the procedure, we can use the fundamental theorem of calculus applied to the pressure:

$$A_N = A_N(1) = A_N(0) + \int_0^1 \frac{dA_N(t)}{dt} dt, \tag{6}$$

By a direct calculation is then trivial to show that the pressure of the ferromagnetic model in Guerra's interpolation scheme is given by

$$\begin{aligned} A_N &= \ln 2 + \ln\left[\cosh\left(\beta \bar{m}\right)\right] - \left(\frac{\beta}{2}\right) \bar{m}^2 + \int_0^1 \left(\frac{\beta}{2}\right) \left\langle (m_N - \bar{m})^2 \right\rangle_t dt \\ &= A^{trial}(\bar{m}) + \int_0^1 R_N(t;\bar{m}) dt, \end{aligned} \tag{7}$$

where $A^{trial}(\bar{m}) = \ln 2 + \ln\left[\cosh\left(\beta \bar{m}\right)\right] - \left(\frac{\beta}{2}\right) \bar{m}^2$. The rest $R_N(t;\bar{m}) = \beta/2 \langle (m_N - \bar{m})^2 \rangle_t$, that one wants to remove or reduce as possible, is positive defined and represents the fluctuation of the magnetization around $\bar{m}$. Since in the thermodynamic limit the magnetization is a self averaging order parameter, it is possible to find an optimum $\bar{m}$ such that $P(m) = \delta(m - \bar{m})$ and consequently $\lim_{N\to\infty} R_N(t;\bar{m}) = 0$. From the positivity of the rest, it is easy to see that the optimum $\bar{m}$ can be found by minimizing the trial free energy $-\beta^{-1} A^{trial}(\bar{m})$. In this way we obtain the self-consistent equation which rules the behavior of the order parameter itself (from the previous considerations we can argue that the optimal trial parameter $\bar{m}$ assumes the physical meaning of the thermodynamic limit of the system's magnetization itself, i.e. $\bar{m} = \lim_{N\to\infty} \langle m_N(\sigma) \rangle := \langle m \rangle$):

$$\langle m \rangle = \tanh\left[\beta(\langle m \rangle + h)\right]. \tag{8}$$

In order to simplify the understanding of the bridges we pursue, it is convenient to plot the behavior of the order parameter versus the two tunable parameters, noise level β and external field h (see Fig. 1).

3 Ferromagnetic behavior in quantitative sociology

In the following we briefly summarize results obtained in the analysis of social networks, particularly focusing (for the sake of concreteness) on immigration phenomena [11], reporting a quantitative result from [12].

In this context we want to show that classical integration quantifiers like the percentage of mixed marriages, once plotted versus the percentage of migrants inside the host community, behaves as the magnetization versus the temperature of classical mean-field ferromagnetism, namely with the order parameter scaling as a square

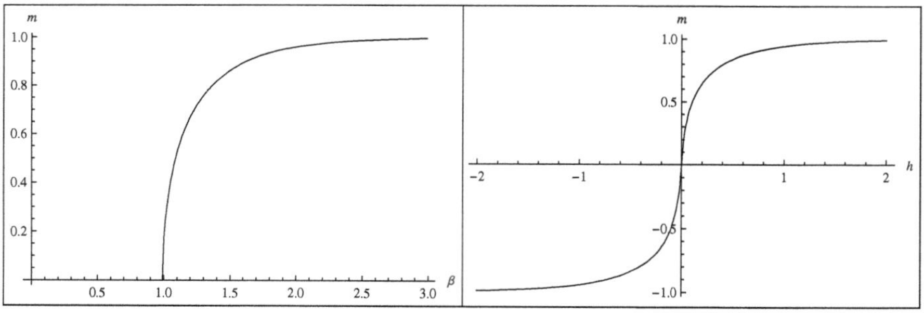

Fig. 1 Left panel: the behavior of the order parameter m as a function of the noise level β can be formally described by a square-root function. Right panel: the behavior of the order parameter m as a function of the external field h can be formally described by a hyperbolic tangent, i.e. a sigmoidal curve

root of the tunable noise level. Calling N_{imm} the amount of immigrants in the host country and the total population (of immigrants and natives) $N = N_{imm} + N_{nat}$ and defining $\gamma = N_{imm}/N$, a natural parameter for assessing change in integration quantifier is the product $\Gamma \equiv \gamma(1-\gamma)$ as

$$N_{imm}N_{nat} \propto \Gamma, \tag{9}$$

since it counts the number of possible cross-group links. By analyzing a database on immigration and integration from Spain in the time window 1990–2000, we found that the quantifier capturing the mixed-marriages displays non-linear behavior, in particular it follows remarkably a square root (see Fig. 2).

Understanding such a behavior from a statistical mechanics perspective is quite simple: let us consider two ensembles of agents, the natives, denoted by $\pm 1 \ni \sigma_i$, $i \in (1, \dots, N)$ and the immigrants, denoted by $\pm 1 \ni \tau_\mu$, $\mu \in (1, \dots, P)$. Of course $\gamma = P/(P+N)$. The values ± 1 coupled to the possible values of σ and τ stand for a positive attitude (+1), or its lack (-1), with respect to contracting a marriage with

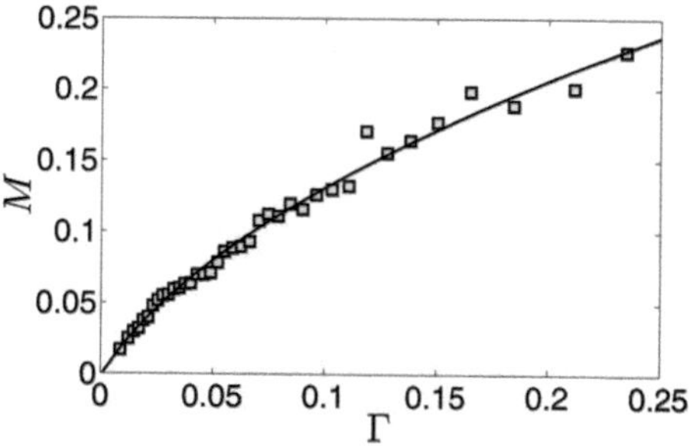

Fig. 2 The plot shows the (normalized) amount of mixed marriages versus the percentage of migrants inside the host country: squares are real data mirroring the decade 1990–2000 in Spain while the continuous black curve is the best fit with a trial $f(\Gamma) = a\sqrt{\Gamma}$ with $a \sim 0.52 \pm 0.02$ and an $R^2 \sim 0.992$. Note that here m plays as the magnetization in the mean-field Ising model while Γ plays as the (rescaled) temperature. Data are from [12]

an immigrant and a native, respectively. If we believe that imitation plays a role in social networks, it is then possible to built an Hamiltonian as

$$H(\sigma,\tau) = \frac{-1}{P+N}\sum_{i,\mu}^{N,P} \sigma_i \tau_\mu, \tag{10}$$

that represents the following: stable (potential) couples are those where the members are both happy or unhappy with the mixed marriage. What is unfavorable is a long-term state where one of the two members wants the mixed marriage but the other does not. We can then built the statistical mechanics machinery to see what this prescription implies. The partition function reads off as

$$Z(\beta,\gamma) = \sum_\sigma \sum_\tau \exp\left(\frac{1}{P+N}\sum_{i,\mu}^{N,P} \sigma_i \tau_\mu\right) \sim \sum_\sigma \exp\left(\frac{\gamma(1-\gamma)}{2N}\sum_{ij} \sigma_i \sigma_j\right), \tag{11}$$

which is nothing but the partition function of a (single party) ferromagnetic model with coupling $J = \gamma(1-\gamma)$.

Following the previous section (as we reduced to that framework) we know how to write the self-consistency, which reads here as

$$\langle m \rangle = \tanh\left[\beta\gamma(1-\gamma)\langle m \rangle\right] \sim \beta\sqrt{\gamma(1-\gamma)}. \tag{12}$$

Hence, if imitation has a key role in social networks we expect that the average attitude of the population versus the percentage of migrants, and, ultimately the number M of mixed marriages, depends on Γ as $M \sim \sqrt{\Gamma}$, exactly as we experimentally found in this test-case (see Fig. 2).

4 Ferromagnetic behaviors in biochemistry

Chemical kinetics usually considers a hosting molecule P that can bind N identical molecules S on its structure; calling P_j the complex of a molecule P with $j \in [0:N]$ molecules attached, the reactions leading to the chemical equilibrium are the following: $S + P_{j-1} \rightleftharpoons P_j$, and, as a convenient experimental observable, usually the average number $\overline{S}$ of substrates bound to the protein is considered as

$$\overline{S} = \frac{\sum_{i=1}^N i[P_i]}{\sum_{i=1}^N [P_i]} = \frac{\sum_{i=1}^N iK^{(i)}[S]^i}{1+K^{(1)}\sum_{i=1}^N K^{(i)}[S]^i}, \tag{13}$$

which is the well-known Adair equation [4], whose normalized expression defines the *saturation function* $Y = \overline{S}/N$. More generally, one can allow for a degree of sequentiality and write

$$\overline{Y} = \frac{K[S]^{n_H}}{1+K[S]^{n_H}}, \tag{14}$$

which is the well-known Hill equation [19], where n_H, referred to as Hill coefficient, represents the effective number of substrates which are interacting, such that

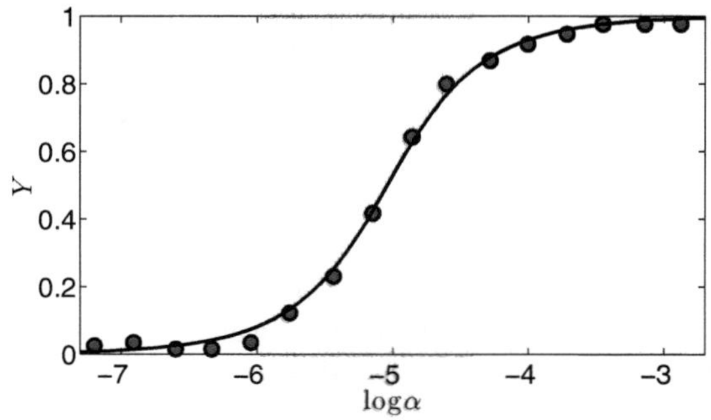

Fig. 3 The plot shows the saturation curve (the normalized amount of bind ligands) versus the log-concentration of the substrate in a simple titration. Real data (∘) representing Ca^{2+}-calmodulin-dependent protein kinase are best fitted (solid line) using the self-consistency 1.16, with $R^2 \sim 0.98$. Note that here Y plays as the (shifted) magnetization of the mean-field Ising model, while $\log(\alpha)$ plays as the external field h. Data are from [4]

for $n_H = 1$ the system is said to be *non-cooperative* and the Michaelis-Menten law is recovered while for $n_H > 1$ it is *cooperative.* In order to bridge this scenario with statistical mechanics, following [4], let us consider an ensemble of elements (e.g. identical macromolecules, homo-allosteric enzymes, etc.), whose interacting sites are overall N and labeled as $i = 1, 2, \dots, N$. Each site can bind one smaller molecule (e.g. of a substrate) and we call α the concentration of the free molecules ($[S]$ in standard chemical kinetics language as used before). We associate to each site an Ising spin such that when the ith site is occupied $\sigma_i = +1$, while when it is empty $\sigma_i = -1$. A configuration of the elements is then specified by the set $\{\sigma\}$.

We model the interaction between the substrate and the binding site by an external field h meant as a proper measure for the concentration of free-ligand molecules, hence $h = h(\alpha)$. We can think at h as the chemical potential for the binding of substrate molecules on sites: when it is positive, molecules tend to bind to diminish energy, while when it is negative, bound molecules tend to leave occupied sites. The chemical potential can be expressed as the logarithm of the concentration of binding molecules and one can assume that the concentration is proportional to the ratio of the probability of having a site occupied with respect to that of having it empty, and we can pose $h = \frac{1}{2}\ln(\alpha)$.

Similarly to the previous test-case drawn from sociology, here we focus again on pairwise interactions and we use complete bipartite graph structure. Sites are divided in two groups, referred to as A and B, whose sizes are N_A and N_B ($N = N_A + N_B$), respectively. Each site in A (B) is linked to all sites in B (A), but no link within the same group is present. As a result, given the parameter J and h, the energy associated to the configuration σ turns out be

$$H_N[\{\sigma\}; J, h] = -\frac{1}{N_A + N_B} \sum_{i=1}^{N_A} \sum_{j=1}^{N_B} J \sigma_i \sigma_j - h \sum_{i=1}^{N_A+N_B} \sigma_i. \tag{15}$$

Note that in (15) the sums run over all the binding sites: despite we deal with the thermodynamic limit, this does not imply that we model macromolecules of infinite length, which is somehow unrealistic, but that we can consider N as the total number of binding sites, localized even on different macromolecules, as boundary effects can be reabsorbed in an effective renormalization of the couplings $J \geq 0$.

A key point is that the saturation function $Y(\alpha)$ is closely related to the magnetization in statistical mechanics $m(h)$ as it reads off as

$$Y(\alpha) = \frac{1}{2}\sum_{i=1}^{N}(1+\sigma_i) = \frac{1}{2}[1+\langle m(h(\alpha))\rangle].$$

Recalling the expression for the self-consistency equation, we are immediately able to see that $Y(\alpha)$ fulfills the following free-energy minimum condition

$$Y(\alpha) = \frac{1}{2}\{1+\tanh[J(2Y-1)+\frac{1}{2}\log(\alpha)]\}. \tag{16}$$

This expression returns the average fraction of occupied sites corresponding to the equilibrium state for the system. In general, the Hill coefficient can be obtained as the slope of $Y(\alpha)$ in Equation (16) at the symmetric point $Y = 1/2$, namely

$$n_H = \frac{1}{Y(1-Y)}\frac{\partial Y}{\partial \alpha}\bigg|_{Y=1/2} = \frac{1}{1-J}. \tag{17}$$

Further, the expression in Equation (16) can be used to fit experimental data for saturation versus substrate concentration. As shown in Figure 3, the fit of experimental data is very good and Hill coefficients derived in this way and the related estimates found in the literature are also in excellent agreement.

5 Ferromagnetic behaviors in electronics

This section is dedicated to the understanding, within the statistical mechanics framework, of collective behaviors in cybernetics; in particular, we focus on the electronic declination of cybernetics because this is probably the most practical and known branch [26].

Following [4], the plan is to compare self-consistencies in statistical mechanics and transfer functions in electronics so to reach a unified description for these systems.

The core of electronics is the operational amplifier, namely a solid-state integrated circuit (transistor), which uses feed-back regulation to set its functions.

An ideal amplifier is the *linear* approximation of the saturable one and essentially assumes that the voltage at the input collectors is always at the same value so that no current flows inside the transistor [26].

If we call V_{out} the output signal (in Volts) and V_{in} the input signal (in Volts) that exits/enters the amplifier, and we call R_2 (in Ω) the resistor that allows for retroaction (feed back signal), then the transfer function of the system can be obtained as

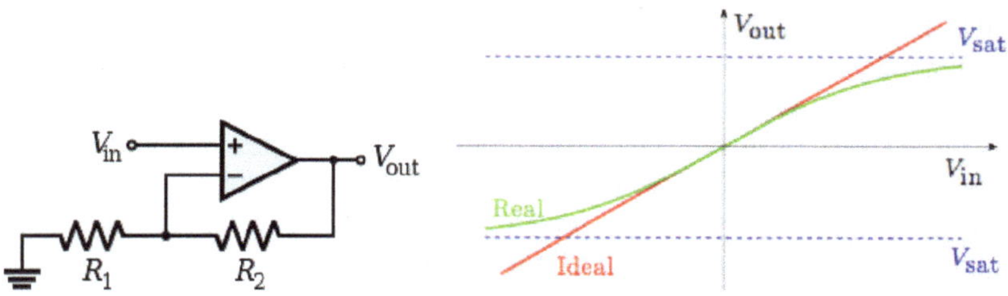

Fig. 4 Left panel: Schematic representation of an operational amplifier. Note that R_2 modulates the strength of the feed-back signal, the retroaction that allows amplification, as J modulates the feed-back of a spin via its nearest neighbors in statistical mechanics. Right panel: Transfer function, namely $V_{out}(V_{in})$. Red line is the "ideal response", where the transistor linearly amplifies, while the green curve represents the real transfer function, showing the collapse to the asymptotes $\pm V_{sat}$, where V_{sat} is the saturation value reached by the amplifier. Note the manifest behavior at the hyperbolic tangent, typical of ferromagnetism. Data are from [4]

$V_{out} = (1+R_2)V_{in}$ [26] (without loss of generality we set $R_1 = 1\Omega$ for the external resistor, see Fig. 4 (left)). Therefore, as far as $R_2 > 0$, the gain is larger than one and the circuit is amplifying the input.

To highlight our parallel, we note that the transfer function is an input/output relation, exactly as the equation for the order parameter m. In fact, for small values of the coupling J (so to mirror ideal amplifier), we can write

$$\langle m \rangle \sim (1+J)h. \tag{18}$$

Thus, the external signal V_{in} is replaced by the external field h, and the response of the system V_{out} is replaced by the response of the system $\langle m \rangle$. By comparison we see that R_2 plays as J, and, consistently, if on the electronic side $R_2 = 0$ the retroaction is lost and the gain is no longer possible: this is perfectly consistent with the statistical mechanics perspective, where if $J = 0$ spins do not mutually interact and no feed-back is allowed. The sigmoidal shape of the hyperbolic tangent is not accounted by ideal amplifiers: this is because saturation is not included in the approximation we discussed, however, it simply makes V_{sat} asymptotes for the growth, hence recovering the expected behavior, as shown in Figure 4 (right).

6 Ferromagnetic behaviors in theoretical immunology

Concerning cooperation in biology, we focus on the field of immunology, as we spent some years studying the emerging collective behavior of lymphocytes, see e.g. [3,5].

The immune system is a marvelous and extremely complex ensemble of different cells and signalling proteins: we will focus only on a sub-shell of the whole system, namely the population of B-cells, the soldiers dedicated to the antibody production. Classical clonal selection theory [25] assumes that a host-body has an enormous amount of different B-cells producing different antibodies. B-cells producing the same antibody are grouped into "clones" and the collection of all the clones forms the "repertoire". Clones have the peculiarity, beyond antigenic recognition, to respond also to stimulation from other lymphocytes: if a clone is releasing antibodies and those are complementary enough to the receptors of another clone, the latter will start to release antibodies as well. This mechanism, called "elicitation" in immunology, strongly resembles imitative behavior and gives rise to the so called Jerne idiotypic network [10], whose properties we want briefly to outline.

Proceeding along a general information theory perspective, we associate to each antibody, labeled as i, a binary string Ψ_i of length L, which effectively carries information on its structure and on its ability to form complexes with other antibodies or antigens. Since antibodies secreted by cells belonging to the same clone share the same structure, the same string Ψ_i is used to encode the specificity of the whole related B clone. In this way, the repertoire will be represented by the set N of properly generated strings. Antibodies can bind each-other through "lock-and-key" interactions, that is, interactions are mainly hydrophobic and electrostatic and chemical affinities range over several orders of magnitude [25]. This suggests that the more complementary two structures are and the more likely (on an exponential scale) their binding. We therefore define χ as a Hamming distance, to measure the complementarity between two bit-strings and introduce a phenomenological coupling

$$\chi_{ij} = \sum_{k=1}^{L} [\Psi_i^k(1-\Psi_j^k) + \Psi_i^k(1-\Psi_j^k)] \Rightarrow J_{ij} \propto e^{\alpha \chi_{ij}} \tag{19}$$

where α tunes the interaction strength.

Hence, different clones interact with external antigens and among each other with a coupling given by their reciprocal binding affinities of the corresponding antibodies. The latter can be formalized in Hamiltonian terms as follows:

$$H_N(\sigma|J) = -\frac{1}{N}\sum_{i<j} J_{ij}\sigma_i\sigma_j - \sum_i h_i\sigma_i, \tag{20}$$

where N is the total number of different clones (the size of the repertoire), the dichotomic spin σ_i may assume values $+1$ representing antibody release or -1 representing quiescence and the positive coupling $J_{ij} \geq 0$ ensures reciprocal elicitation (imitation) when different from zero, and h_i represents the antigenic load (implying a response by node i).

Now a crucial observable is the weighted connectivity, defined as $W_i = \sum_j^N J_{ij}$, whose distribution $P(W)$, exploiting the fact that couplings are log-normally distributed (see Eq. (19) and [6, 10]), can be approximated as

$$P(W) \sim \frac{1}{W\tilde{\sigma}\sqrt{2\pi}} e^{-\frac{(\ln(W)-\tilde{\mu})^2}{2\tilde{\sigma}^2}}, \tag{21}$$

where $\tilde{\mu}$ and $\tilde{\sigma}$ are related to the J_{ij} distribution (a detailed derivation of these values can be found in [6]). The last observable deserves attention as it can be compared with experimental results performed on ELISA technology on mice, as reported in Figure 5, depicting data from [6]: again, there is a remarkable agreement between real data and ferromagnetic predictions.

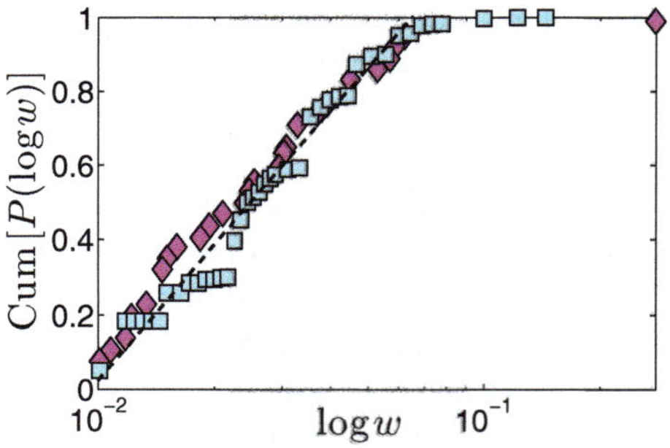

Fig. 5 Plot shows the cumulative distribution (frequency for the experimental part) of the weighted connectivity shown by B-lymphocytes in mice. Data are elaborated from [6]: Rumbles are experimental data, squares are Monte Carlo simulations with our $P(W)$ and the dashed black line represents the theoretical scaling of $P(W)$

7 Conclusions

In view of broad applications, suitable to help the scientific community in properly framing complexity of Planet Earth into major scaffolds, in these notes we revised the paradigmatic ferromagnetic mean-field scenario, embedding positive coupling among spins on a random graph, such that nodes represent spins and links, whenever present, mirror their interactions.

Beyond a classical role in depicting the essence of phase transitions and spontaneous symmetry breaking in theoretical physics, this model, and more properly the statistical mechanics approach to model cooperativity, is finding a renewed role in tackling the emergent behavior of disparate systems as a function of tunable external parameters. Indeed, applications have focused on a broad range of systems, all sharing the same microscopic structure, made of by several interacting elements (theoretically denoted as "spins" and whose nature is specified by the particular example considered) via positive (imitative) couplings.

In these notes we showed, trough several examples and comparisons with real data, that in chemistry (with the example of reaction kinetics, where spins are ligands), in biology (stemming from the idiotypic network of lymphocytes where spins are B-clones), in sociology (by investigating migrant's integration inside a host community where spins are the decision makers) and in electronics (analyzing the transfer function of operational amplifiers, where spins are internal junctions), the statistical mechanics of ferromagnetism is able to properly describe the complex, emergent, phenomenology of their order parameters.

Hence, the role of this review is to highlight a key, unifying, role performed by this technique in showing that systems apparently diverse and unrelated, behave in the same way once properly described. We believe that merging separate disciplines by finding a "universal behavior" is an important requisite in order to quantify the complexity of such fields, which, ultimately, reflects the complexity of Planet Earth, focus of the present volume.

Acknowledgements The authors acknowledge the FIRB grant RBFR08EKEV, Sapienza Universitá di Roma, INFN and INdAM for financial support.

Our colleagues, alphabetically ordered as Raffaella Burioni, Pierluigi Contucci, Gino Del Ferraro, Aldo Di Biasio, Francesco Guerra, Francesco Moauro, Richard Sandell, Guido Uguzzoni and Cecilia Vernia, are truly acknowledged for walking together into this fascinating research route.

References

1. Acebron, J.A. et al: The Kuramoto model: A simple paradigm for synchronization phenomena. Rev. Mod. Phys. **77**(1), 137 (2005)
2. Agliari, E., Barra, A.: A Hebbian approach to complex-network generation. Euro Phys. Lett. **94**(1), 10002 (2011)
3. Agliari, E., Barra, A., Guerra, F., Moauro, F.: A thermodynamical perspective of immune capabilities. J. Theor. Biol. **287**, 48–63 (2011)
4. Agliari, E., Barra, A., Burioni, R., Di Biasio, A., Uguzzoni, G.: Biochemical kinetics and cybernetics. Submitted to Scientific Reports, Nature Publishing Group (2013)
5. Agliari, E., Barra, A., Galluzzi, A., Guerra, F., Moauro, F.: Multitasking associative networks. Phys. Rev. Lett. **109**(26), 268101 (2012)
6. Agliari, E., Barra, A., Del Ferraro, G., Guerra, F., Tantari, D.: Anergy in self-directed B lymphocytes: A statistical mechanics perspective. Submitted to Scientific Reports. Nature (2013)
7. Ashcroft, N.W., Mermin, N.D.: Solid state physics. Dover Press, Baltimore (1976)
8. Barra, A.: The mean field Ising model trough interpolating techniques. J. Stat. Phys. **132**(5), 787–809 (2008)
9. Barra, A., Agliari, E.: Equilibrium statistical mechanics on correlated random graphs. JSTAT P02027 (2011)
10. Barra, A., Agliari, E.: A statistical mechanics approach to autopoietic immune networks. JSTAT P07004 (2010)
11. Barra, A., Contucci, P.: Toward a quantitative appproach to migrant's integration, Eur. Phys. Lett. **89**(6), 68001 (2010)
12. Barra, A., Contucci, P., Sandell, R., Vernia, C.: Integration indicators in immigration phenomena: A statistical mechanics perspective. Submitted to Scientific Reports, Nature (2013)
13. Barra, A., Genovese, G., Guerra, F.: Equilibrium statistical mechanics of bipartite spin systems. J. Phys. A **44**(24), 245002 (2011)
14. Barra, A., Genovese, G., Guerra, F., Tantari, D.: How glassy are neural networks? JSTAT 07, 07009 (2012)
15. Baxter, R.J.: Exactly solved models in statistical mechanics. Academic Press, Canberra (2007)
16. Benson, S.W., Benson, S.W. : The foundations of chemical kinetics. McGraw-Hill, New York (1960)
17. Bollobas, B.: Random graphs. Cambridge University Press, London (2001)
18. De Sanctis, L., Guerra, F.: Mean field dilute ferromagnet: high temperature and zero temperature behavior. J. Stat. Phys. **132**(5), 759–785 (2008)
19. Di Biasio, A. et al: Mean-field cooperativity in chemical kinetics. Theor. Chem. Accounts **131**(3), 1–14 (2012)
20. Durlauf, S.N.: How can statistical mechanics contribute to social science? Proc. Natl. Acad. Sc. USA **96**(19), 10582–10584 (1999)
21. Ellis, R.S.: Entropy, large deviations, and statistical mechanics. Taylor and Francis, New York (2005)
22. Evans, D.J., Morriss, G.: Statistical mechanics of non-equilibrium liquids. Cambridge University Press, Cambridge (2008)
23. Gallo, I., Contucci, P.: Bipartite mean field spin systems. Existence and solution. Math. Phys. E. J. **14**, 463 (2008)

24. Guerra, F.: An introduction to mean field spin glass theory: methods and results. Lectures at the Les Houches Winter School (2005)
25. Janaway, C.A. et al: Immunobiology. Garland Science Press, New York (2003)
26. Millman, J., Halkias, C.C.: Integrated electronics: Analog and digital circuits and systems. Allied Publishers, Philadephia (1972)
27. Onsager, L.: The Ising model in two dimensions. In: Critical Phenomena in Alloys, Magnets and Superconductors, pp. 3–12. McGraw-Hill, New York (1971)
28. Thompson, C.J.: Mathematical statistical mechanics. The MacMillan Company, Toronto (Ontario) (1976)

The Near Earth Asteroid Hazard and Mitigation

Ettore Perozzi

Abstract Since the 1898 discovery of Eros, the first asteroid known to approach our planet, the so-called Near-Earth Asteroid population has grown to exceed at present 10000 objects. As such it represents a threat for our planet as well as an opportunity for science and exploration. An introduction to asteroid hazard and mitigation issues is presented, together with basic mission design considerations.

1 Introduction

The Near Earth Asteroid (NEA) population is composed of celestial bodies of relatively small size (rarely exceeding 10 km) whose orbital characteristics allow to closely approach our planet. As such they are at risk of collision with the Earth and much effort has been devoted in the last 20 years to evaluate the impact hazard and to develop mitigation strategies. NEAs are also interesting for science because of their link with meteorites, as primitive bodies hosting pristine samples of the raw material from which the Solar System formed. In recent times NEAs have gained increasing attention as potential extraterrestrial resources and as targets for the next step of human exploration. In this respect the asteroid retrieval mission proposal, which aims to bring a small asteroid in a geocentric orbit farther than the Moon could fulfil both objectives. Within this framework, mathematics has played a crucial role, from modelling asteroid encounters and the associated uncertainties to evaluating the accessibility of the NEA population for space mission design. In what follows reference to the original techniques and the results achieved is given and some simple cases are discussed in more detail.

E. Perozzi (✉)
Elecnor Deimos, Ronda de Poniente 19, 28760 Tres Cantos, Madrid, Spain
IAPS-INAF, via Fosso del Cavaliere 100, 00133 Roma, Italy
e-mail: ettore.perozzi@deimos-space.com

A. Celletti, U. Locatelli, T. Ruggeri and E. Strickland (eds.): *Mathematical Models and Methods for Planet Earth.* Springer INdAM Series 6, DOI 10.1007/978-3-319-02657-2_7,

2 Celestial mechanics and close encounters

The very first observation of a celestial body undergoing close gravitational interaction was not involving an asteroid, but a comet and specifically the 1770 appearance of comet Lexell. Yet, as described in [23], this prompted the need for treating differently the whole orbit determination problem and led to the development of new mathematical models for dealing with it, thus paving the way for modern NEA impact monitoring.

Comet Lexell was in fact discovered after a very deep encounter with Jupiter which lowered its perihelion to the point of becoming an Earth crosser, while the fact that after 1779 it was not seen again was due to a subsequent encounter with the same planet which ejected it on a much larger ellipse. Yet the uncertainties with which the comet orbit was known allowed multiple solutions due to the peculiarity of the event involved, to which astronomers and celestial mechanicians were confronted for the first time. The practice of dealing at once not only with an orbital path but also with its possible variations through a parametrization process consistent with the observations represented a breakthrough for properly facing trajectories undergoing strong gravitational interactions. This approach is nowadays extensively applied to NEAs and it is at the core of state-of-the-art impact monitoring techniques. Each time a new object is discovered, the whole region of confidence associated to the nominal trajectory is analysed and the so-called VA (Virtual Asteroid) trajectories are propagated forward in time searching for impactors. This is not a trivial task if one wants to ensure the completeness of the solutions found, in particular when the identification of VIs (Virtual Impactors, i.e. collision solutions) are found. The widely different dynamical characteristics of the NEAs, the large uncertainties often associated with discovery observations and the chaotic behaviour resulting from close encounters, have required to considerably extend the domain of applicability of standard mathematical methods such as the least squares fit for orbit determination, the LOV (line of variation) sampling and the algorithms for searching minimum distance at close approach [14]. Also classical celestial mechanics issues such as Opiks theory of interplanetary encounters [17], MOID – Minimum Orbit Intersect Distance – computation [10, 11] and resonances exploitation through resonant returns [6, 24], enter the problem and have undergone substantial developments.

As an example of this attitude it is then worthwhile discussing how the Tisserand quantity T, which was originally developed for the identification of comets, has assumed a novel significance when applied to the different classes of small bodies of the Solar System, as well for quickly providing some basic parameters when dealing with close encounters. Derived from the general formulation of the Jacobi integral in the three-body problem, the Tisserand quantity does not change when the small body is far from the two primaries and can be espressed by knowing the semimajor axis a, the eccentricity e and the inclination i of the small body orbit:

$$T = \frac{1}{a} + 2\sqrt{a(1-e^2)}\cos i$$

where the semimajor axis is measured in units of that of the primaries. Being an invariant of the Sun-Jupiter-small body dynamics, this quantity represented a powerful mean for astronomers to check by very simple computation if two comets exhibiting different orbital parameters were in fact the same object reappearing after a close encounter with Jupiter, whatever deep.

A broader meaning to the Jacobi/Tisserand quantity is due to the seminal work of Kresak [12], who pointed out that it could be used for dynamically characterising the various populations of small bodies of the Solar System and their interrelations. T is in fact related to the planetocentric velocity at encounter U by:

$$U = \sqrt{3 - T}$$

measured in units of the planet velocity. Thus it is possible to clearly separate, for example, Main-Belt Asteroids ($T > 3$) from NEAs. For the latter, one can also compute rather straightforwardly the Earth flyby (or impact) velocity V by taking into account the gravitational focussing of our planet:

$$V = \sqrt{11.2^2 + U^2}$$

by going back to traditional units and introducing the Earth escape velocity of 11.2 km/s. As it will be shown below, this result can be used also for space mission analysis and in particular for performing target selection for a NEA sample return mission.

3 Space Situational Awareness

The Earth, like the other terrestrial planets, is exposed to the NEA hazard as a result of the chaotic diffusion toward the inner solar system of asteroidal fragments generated by major collisions in the main asteroid belt, located between the orbits of Mars anf Jupiter. These fragments can be as small as dust grains thus completely burning in the atmosphere or, in order of increasing size from meters to km-size objects, landing on our planet as small meteorites, exploding catastrophically in the atmosphere producing bright fireballs or hitting the surface at velocities high enough to cause local or global damage.

Yet the Earth, unlike the other terrestrial planets, is inhabited by a human civilization increasingly dependent from technology (e.g. power grids, artificial satellites, high-speed communication networks). Therefore protecting the planet from cosmic hazards means not only to be able to timely discover "dinosaur-killer" asteroids, i.e. those large enough to threaten the survival of humankind, but also the much smaller ones which are nevertheless potentially capable to produce direct or indirect casualties associated to their fall. The corresponding mitigation actions can vary from sending an interceptor into space to deflect the asteroid from its collision course with our planet, to the evacuation of the interested region and/or to the implementation of safety measures for citizens and sensible infrastructures. In the recent Chelyabinsk superbolide event all injuries were due to the side effects of the shockwave produced by the in-flight explosion of the meteoroid and could have been certainly avoided if the parent impactor were detected a couple of days in advance by simply instructing the population on how to properly behave during the event.

Solving the NEA risk problem means to address with a high level of reliability and completeness a number of logical steps; in all of them mathematics plays a prominent role. Discovering hazardous objects comes first and the sharp growth in the number of known NEAs at the beginning of this century witnesses the success of the introduction of wide field sky surveys (Fig. 1). In the last ten years were found ten times more NEAs than in the whole century starting with the discovery of Eros, the first NEA, in 1898. Although this huge effort has led to the detection of almost all NEAs larger than 1 km (i.e. those which are likely to produce global catastrophes leading to mass extinctions), only 10% of the hundred meter-size objects are known to date. When dealing with the even smaller ones, yet still able to produce substantial damage (the Chelyabinsk meteoroid estimate is of slightly less than 20 meters), then the figure drops well below 1%. In this respect it is important to point out that with modern CCD technology, the ability to detect a faint moving object in the sky depends not only on the characteristics of the telescope(s) used but also on the data processing algorithms employed. A successful NEA sky survey must be capable of efficiently and autonomously inspecting all images produced by a telescope network each night (which can be of the order of thousands for the next generation sky surveys) avoiding false detections and correctly correlating the point-like sources belonging to the same object.

Once an object is detected orbit determination is needed to check whether the object is a new discovery and if so impact monitoring is triggered. Both functions make extensive use of the mathematical methods mentioned in the previous section for dealing with planetary close encounters. Determining if and when the Earth could be hit by a wandering asteroid and estimating the corresponding probability is at the heart of the NEA risk problem. There are at present two software robots which address this issue: CLOMON2 [15] and Sentry [4]. Both produce a list of objects which have a non-negligible impact probability. This step in turn, leads to close the "risk problem" loop, because it allows to rank the danger associated to each asteroid, thus indicating to the observers which objects deserve urgent follow-up observations in order to decrease the orbital uncertainty. An updated orbit determination and impact monitoring loop is then performed. The iterations usually end when the orbit improvement allows to exclude all impact solutions.

An intermediate case occurring rather frequently happens when the follow-up observations are not deciding and the object becomes quickly unobservable, thus leaving open the impact assessment at least until its next apparition. The only predicted impact to date, is that of 2008TC3, a 5-meter size meteoroid which produced a meteorite fall in the Sudan desert.

At present two operational systems have been designed to receive as an input the observations performed worldwide, carrying out routinely and with a high degree of automation all the steps of the risk problem just described. These are NEODyS (Near-Earth Objects Dynamic Site – http://newton.dm.unipi.it/neodys) developed at the University of Pisa, and the Near Earth Object Program (http://neo.jpl.nasa.gov)

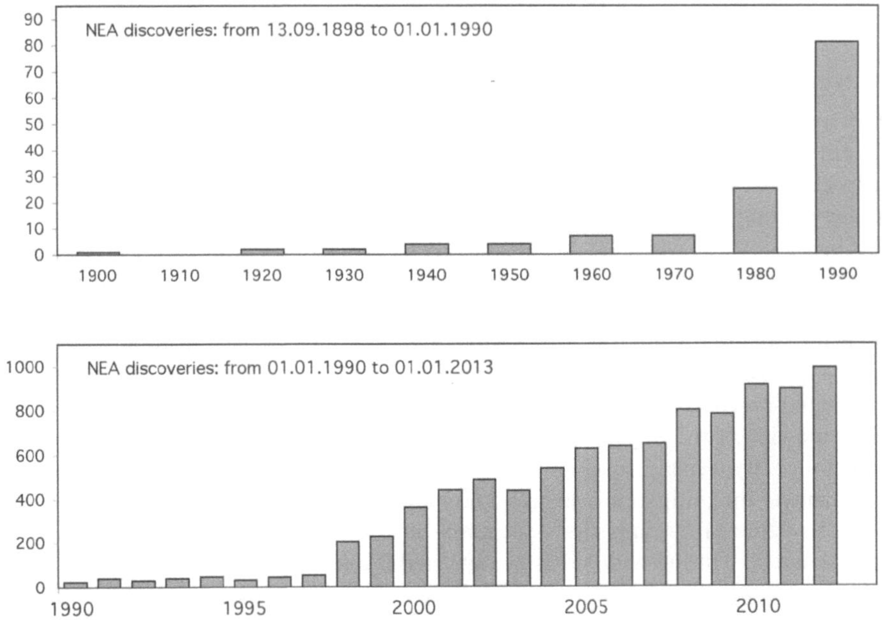

Fig. 1 Histogram of the distribution of NEA discoveries in time. In 2012 the discovery rate raised to almost 1000 objects per year (Courtesy ESA NEO Coordination Centre)

running at JPL. They use different methods for performing impact monitoring and keep a strong link during operations thus providing the necessary redundancy for dealing with stiff problems, such as those associated with the Apophis case [5], the asteroid which in December 2004 scored the highest impact probability ever.

Within this framework Europe has recently initiated the Space Situational Awareness Programme whose aim is *to support the European independent utilisation of and access to space for research or services, through providing timely and quality data, information, services and knowledge regarding the environment, the threats and the sustainable exploitation of the outer space surrounding our planet* (cfr. the ESA SSA Programme Declaration). The programme is structured into three Segments addressing Space Debris (SST), Space Weather (SWE) and Near-Earth Objects (NEO). The NEO Segment [7] has been designed to ensure a significant European contribution in the worldwide NEO hazard monitoring scenario, by focussing on the discovery of small NEAs approaching the Earth [9] and on fostering the European excellence in NEO impact monitoring and physical observations [18]. The deployment of the European NEO Coordination Centre has been successfully achieved and precursor services are currently on-going (http://neo.ssa.esa.int).

4 Accessibility and Mitigation

The NEA risk problem cannot be considered completely addressed until two further steps are discussed, closely related to the feasibility of mitigation actions. In principle mitigation can be fully effective in that timely changing the impactor trajectory in space could remove entirely the threat. Unfortunately the effectiveness of such a strategy strongly depends from the warning time, i.e. the time elapsed between impact prediction and when the impact actually occurs. If the warning time is too short there is no possibility of preparing, launching and navigating a spacecraft toward the asteroid for implementing whatever deflection strategy is foreseen (e.g. kinetic impactor, gravity tractor, ion beam). Moreover, as shown by [3], the impulse needed for last-minute deflection could easily assume unrealistic values.

Ground-based mitigation can only aim at damage reduction and this applies also when the size of the impactor is likely to produce local damages to a level which does not justify a space mission in terms of cost effectiveness.

Whatever the mitigation scenario, knowing the size, the composition and the internal structure of the impactor is essential in evaluating how effective mitigation actions can be. Hitting with a massive spacecraft a loosely bound rubble pile celestial body has widely different consequences on its trajectory than having to deal with a monolitic structure. The same applies when an asteroid enters the atmosphere. A one-meter metallic object is likely to reach the ground escavating an impact crater tens of meters wide, while a rocky body with a porous structure will possibly produce only a spectacular fireball. This is why physical observations of asteroids are important for mitigation. Photometry allows to obtain indication on the object size, shape and rotation state, while spectroscopy and polarimetry are used to identify composition and surface features. But of course only the direct exploration by spacecraft, and in particular sample return missions, are able to provide the information needed to model with sufficient accuracy the consequences of an impact on Earth or to quantify the trajectory changes needed for a successful deflection attempt.

In this respect the Japanese Hayabusa spacecraft, notwithstanding a malfunction of the sampling mechanism, has proven that NEA sample return missions are feasible at a reasonable cost. The US Osiris-REX mission, presently under realization phase, and the European Marco Polo mission, still under definition, both plan to return asteroidal samples back to Earth within the next decade. The basics of a kinetic impactor mission demonstrator have been laid by the Don Quijote concept, ranked as top priority by the Near-Earth Object Mission Advisory Panel [8]. In this scenario an orbiting spacecraft is sent to the target asteroid, to be used for both, properly characterize the object before another independent spacecraft is sent for impacting the body, and measure with the desired accuracy the tiny deviation imparted.

In general mitigation can be considered as a specific case of the more general problem of the accessibility of the NEA population. The wide range of orbital characteristics exhibited by NEAs, from low inclination almost circular 1-AU Earth-like orbits to highly elongated, high inclination ellipses, translate often into demanding mission requirements. Depending on the specific case, a NEO can be more accessi-

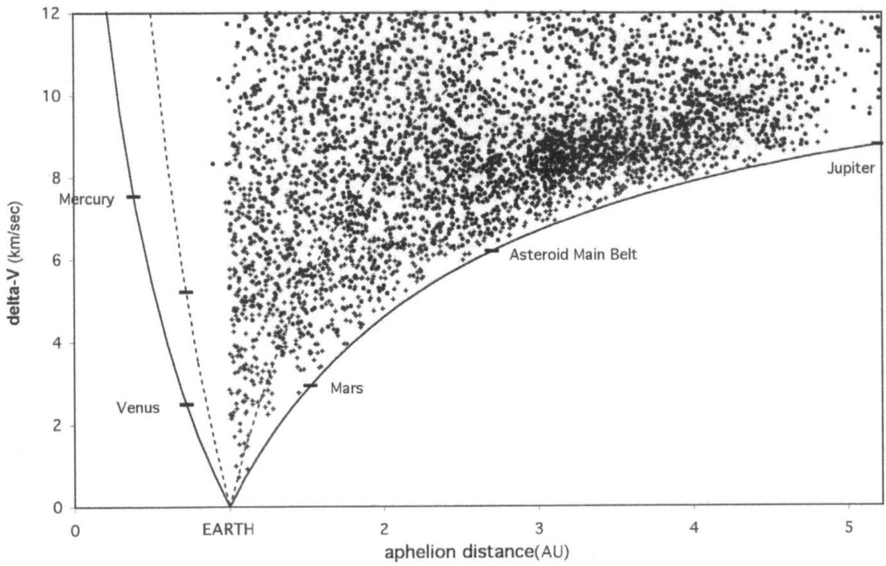

Fig. 2 The distribution of NEAs in the H-plot clearly shows that they are widely dispersed in terms of delta-V. For comparison rendezvous missions to the planets are located on the upper dashed line at the corresponding distances (1.5 AU for Mars, around 2.8 for the asteroid belt, an so on). Objects with absolute magnitude less or equal to 22 (roughly corresponding to 100m in size) are indicated by full circles while for smaller objects crosses are used instead

bile than the Moon [20] or more energy demanding than sending an orbiter around Jupiter or Saturn.

The situation is shown in Figure 2, where the accessibility of the NEA population is represented using an H-plot diagram [19], which displays the total velocity change (delta-V) needed to transform an initially zero inclination circular 1 AU orbit into one identical to that of each object, as a function of its aphelion distance. Should an asteroid have an eccentric coplanar orbit tangent at perihelion or aphelion to that of the Earth, the corresponding point would be located along the lower thick lines of Figure 2. Note that this latter orbit is identical to the intermediate Hohmann transfer trajectory to that distance, i.e the basic astrodynamics tool originally developed to compute the accessibility of the planets (e.g. [22]). Thus any displacement from the lower thick lines is a measure of the additional energy needed to lower or rise the perihelia, as well as changing the inclination of the asteroid orbit. The upper dashed line gives the delta-V needed to perform Hohmann transfers from the Earth to circular orbits of increasing size, thus encompassing those of the planets.

Depending upon the mission requirements (e.g. science, mitigation, technology) the H-plot can be used as a pre-optimization method for target selection because it gives a quick, first-order estimate of the accessibility of candidate targets. As an example, relying on this representation it has been possible to develop an innovative design for a NEA sample return mission [16], where a "free ride" is given by an Earth

Table 1 List of NEO mission target candidates with high Tisserand invariant (T). For each object the following quantities are also reported: the Hohmann transfer delta-V (see text), the catalogue number (if any, in brackets) , the asteroid name (or IAU provisional designation), the absolute magnitude H (related to the size), the spectral type (related to the composition) and the Earth encounter dates. For reference, H = 16 corresponds to km-size bodies, while H = 22 to 100 m class objects. S-type asteroids have a stony composition, C-types are carbonaceous more primitive bodies

Delta-V (km/s)	*(number) name*	*H*	*type*	*Earth encounters*	*T*
3,359	(99942) Apophis	19.2		2021, 29	2.967
3,778	2000 EA14	21.0		2025	2.961
4,221	2002 NV16	21.4		2024	2.974
4,241	(25143) Itokawa	19.2	S		2.964
4,294	(89136) 2001 US16	20.2			2.989
4,407	2003 GA	21.2			2.998
4,692	2001 QC34	20.2		2014, 20	2.961
4,799	2006 QQ23	19.6		2014, 19, 24	2.960
4,803	1999 JU3	19.4	C	2020	2.971
4,867	(65717) 1993 BX3	21.0		2021	2.981
4,977	2006 SU49	19.6		2023, 29	2.964
5,111	2002 SR	21.7		2026	2.963
5,124	1991 JW	19.6		2027, 28	2.964
5,144	2002 TD60	19.3			2.999
5,229	1994 CJ1	21.4			2.978
5,258	2000 FJ10	21.3			2.981
5,277	2003 CC	20.2			2.980
5,318	2003 SD220	16.8		2015, 18, 21, 24, 27	2.967
5,347	2002 JX8	20.4		2023	2.965
5,437	1997 WB21	20.3		2020, 27	2.973

approaching asteroid for delivering the sample back to Earth. In this way the delta-V needed for manoeuvring through the return leg of a traditional sample return mission is not entering the energy budget of the mission. Target selection was performed by checking up objects which were undergoing a close encounter with the Earth in the desired time frame at a relative velocity (computed through the Tisserand quantity T introduced in the previous sections) not exceeding the present technological limit for the survival of a re-entry capsule in the atmosphere (12–13 km/s), and having realistic delta-V requirements. Results are reported in Table 1.

Out of the 20 targets selected, only 2 have an already determined spectral type (which gives an indication of the asteroid composition), while 4 undergo a close encounter with the Earth below 0.05 AU within the timespan 2015–2020.

Note that some objects display encounters with the Earth repeating at equal intervals of time, as in the case of 2006 QQ23 which, upon closer investigation, has been found to be inside the 3:4 mean motion resonance with our planet. From a mission analysis point of view this means that any given mission opportunity is periodically repeated, thus translating into a dynamical behavior well known to celestial mechanics [21].

It is also interesting to remark that the lower V-shaped part of the plot, where the most accessible objects reside, is dominated by small asteroids (black dots). This can be explained by the increasing performances of NEA sky surveys and by their continued operation, which increase the chances to discover fainter objects, and in particular those in the vicinity of the Earth in unfavorable observing geometries (e.g. remaining for a long time at low solar elongations). If until recently such objects were considered not relevant for science and mitigation in space because visiting or deflecting a 10-m asteroid is not cost-effective, the re-assessment of the strategic plans of the major space agencies has brought them on the front line. In particular the NASA plans to choose a NEA as the next step for the human exploration beyond the Moon [1], and for the asteroid retrieval project [13], has re-opened target selection to this class of objects. An extreme accessibility could in fact drastically cut the travel time for humans on a spaceship flying outside the shielding action from cosmic radiation of the Earth magnetic field. Similarly, the rather complex, technologically challenging (e.g. de-spinning the asteroid) and rather expensive in terms of delta-V (consistent changes of the asteroid orbit) steps needed to transfer an asteroid in a stable orbit in the vicinity of the Earth, call for small (less than 10m) targets to become rather appealing.

5 Conclusions

The contribution of mathematics in assessing the asteroid hazard has been summarised, showing that it enters in all the steps needed to address the NEA risk problem. In Figure 3 a tentative estimate of our present understanding of the frequency of occurrence of cosmic impacts on our planet is shown, taking into account the large uncertainties in the meteoroid size distribution, density distribution and in the impactors composition.

The contribution of mathematical models and techniques is expected to further increase in the near future, with the collaborative operation of the new generation of ground based networks of wide field high sensitivity telescopes, which will scan several times per night the entire visible sky. The number of NEA discoveries is likely to grow accordingly thus putting into tight time constraints the impact monitoring and early warning systems. Additionally data fusion techniques are needed to profit of space based assets, even if their primary goals is not NEO observations (e.g. the European Gaia mission). Finally, extensive mission analysis exploiting advanced celestial mechanics is needed for realising both, the challenging next step of the direct explorations of a near Earth Asteroid (either manned or unmanned) and to detect from space impactors coming from nearly the direction of the Sun, which cannot be detected from the ground [25], as learned from the Chelyabinsk superbolide event.

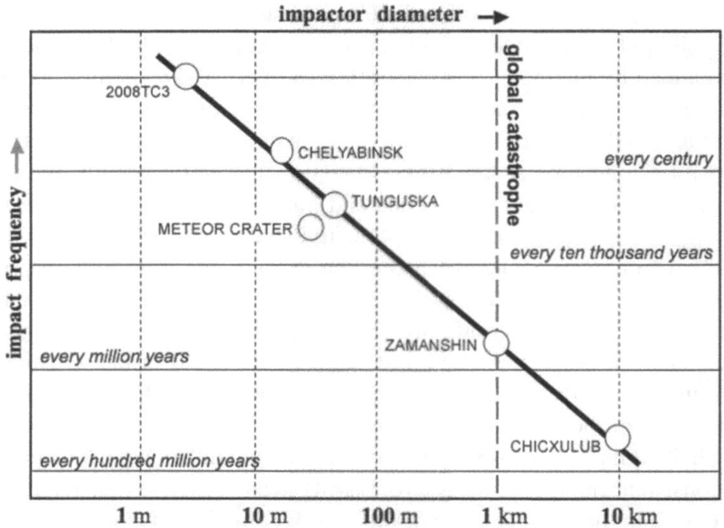

Fig. 3 Impact frequencies and their severity can be extrapolated by known impact cratering and superbolide events. (Courtesy ESA NEO Coordination Centre)

Acknowledgements This work was partially funded under ESA contract No. 4000107291/13/D/MRP – SSA NEO segment precursor service operations (SN-V)

References

1. Abell, P-A. et al: The Near-Earth Object human space flight accessible targets stydy (NHATS) list of of near-earth asteroids: identifying potential targets for future exploration. NASA TR 20120001818 (2012). http://neo.jpl.nasa.gov/nhats/
2. Carusi, A., Perozzi E., Scholl H.: Mitigation Strategy. Comptes Rendus Physique **6**, 367–374 (2005)
3. Carusi, A., Valsecchi, G.B., D'Abramo, G., Boattini, A.: Deflecting NEOs in route of collision with the Earth. Icarus **159**, 417–422 (2002)
4. Chamberlin, A.B., Chesley, S.R., Chodas, P.W., Giorgini, J.D., Keesey, M.S., Wimberly, R.N. Yeomas, D.K.: Sentry: an automated close approach monitoring system for near-Earth objects. Bull. Am. Astron Soc. **33**, 1116 (2001)
5. Chesley, S.R.: Potential impact detection for near-Earth asteroids: The case of 99942 Apophis (2004MN). In: Ferraz Mello, S., Lazzaro, D., Fernandez, J.A. (eds.) Asteroids, Comets, Meteors, Proceedings IAU Symposium No. 229 (2005)
6. Chodas, P.W.: Orbit uncertainties, Keyholes and collision probabilities. Bull. Astron. Soc. Vol. **31**, 1117 (1999)
7. Drolshagen, G., Koschny, D., Bobrinsky, N.: The Near Earth Objects Segment of the European Space Situational Awareness Programme. In: Proceedings IAA Planetary Defense Conference (2011)
8. ESA 2004. Space Mission Priorities for Near-Earth Object Risk Assessment and Reduction, recommendations to ESA by the Near-Earth Object Mission Advisory Panel (NEOMAP)

9. Farnocchia, D., Bernardi, F., Valsecchi, G.B.: Efficiency of a wide-area survey in achieving short- and long-term warning for small impactors. Icarus **219**, 41–47 (2012)
10. Gronchi, G.F.: An algebraic method to compute the critical points of the distance function between two keplerian orbits. Cel. Mech..Dyn. Ast. **93**, 1913–1918 (2005)
11. Gronchi, G.F., Valsecchi, G.B.: On the possible values of the orbit distance between a near-Earth asteroid and the Earth. Mon. Not. Royal Astron. Soc. **492**, 2687–2699 (2013)
12. Kresak, L.: Jacobian integral as a classificational and evolutionary parameter of interplanetary bodies. Bull. Astron. Inst. Slovak. Academy of Sciences **23**(1) (1972)
13. KECK Institute for Space Stidies. Asteroid Retrieval feasibility Study (2012)
14. Milani, A., Chesley, S.R., Boattini, A., Valsecchi, G.B.: Virtual impactors: search and destroy. Icarus **173**, 362–284 (2000)
15. Milani, A., Chesley, S.R., Sansaturio, M.E., Tommei, G., Valsecchi, G.B.: Nonlinear impact monitoring: line of variation searches for impactors. Icarus **173**, 362–284 (2005)
16. NEXT Mission Proposal (2007). TAS-I SD-TN-AI-1067
17. Opik, E.J.: Interplanetary Encounters. Elsevier, New York (1976)
18. Perozzi, E., Bassano, E., Gloria, M., Pagano, F., Reboa, L., Milani, A., Bernardi, F., Farnocchia, D., Valsecchi, G.B., D'Abramo, G., Franco, R., Drolshagen, G., Koschny, D.: Designing the Space Situational Awareness NEO Segment. IAA Planetary Defense Conference: From Threat to Action (2011)
19. Perozzi, E., Rossi, A., Valsecchi, G.B.: Basic Targeting Strategies for Rendezvous and Flyby Missions to the Near-Earth Asteroids. Planetary and Space Science **49**, 3–22 (2001)
20. Perozzi, E., Binzel, R.P., Rossi, A., Valsecchi, G.B.: Asteroids more accessible than the Moon. EPSC2010–750 (2010)
21. Perozzi, E., Casalino, L., Colasurdo, G., Rossi, A., Valsecchi, G.B.: Resonant Flyby Missions to Near-Earth Asteroids. Cel. Mech. Dyn. Astron. **83**, 49–62 (2002)
22. Roy, A.E.: Orbital Motion. Adam Hilger Ltd, Bristol (1988)
23. Valsecchi, G.B.: 236 years ago... In proc.: Milani, A., Valsecchi, G.B., Vokrohulicky, D. (eds.) Near Earth Objects, our celestial neighbours: opportunity and risk. IAU (2007). doi: 10.1017/S1743921307002980
24. Valsecchi, G.B., Milani, A., Gronchi, G.F., Chesley, S.R.: Resonant returns to close approaches: analythical theory. Astron. Astrophys. **408**, 1179–1196 (2003)
25. Valsecchi, G.B., Perozzi, E., Rossi, A.: A space mission to detect imminent Earth impactors. IAU general Assembly abstract 8258 (2012)

Mathematical Models of Textual Data: A Short Review

Mirko Degli Esposti

Abstract This contribution is the result of trying to put in a more systematic, updated and readable form some notes I used in the preparation of my talk at the INdAM Workshop *Mathematical Models and Methods for Planet Earth*, in Rome (May, 2013). The aim was to discuss some recent mathematical approaches to textual data analysis, focusing on literary texts and on some specific topics and examples: universal statistical properties of written language and the nature of long correlations in literary texts with specific applications to authorship attribution, keyword extraction, and automatic text generation.

1 Introduction

Here I narrow my attention to three specific issues and their recent developments concerning mathematical approaches to textual data analysis: in section 1, I tried to summarise some new discussions on the meaning and origin of the old and well known empirical Zipf's law for words [37]. This is a huge argument with several different aspects and I do not pretend to be either complete or original. This section is fully based on the cited papers with the hope to stimulate their reading and some curiosity. There are no personal contributions to the subject, besides the freedom of putting some material and references together. In Section 2, I shift the attention from frequencies of words (invariant by any shuffling of the text) to their (returning time) distribution and long range correlations. While my motivations and interests come from the recent paper in collaborations with E. Altman and G. Cristadoro [2], here I mostly concentrate on the important approach of Montemurro and Zanette to the quantification and extraction of semantic information based on distribution of words

M. Degli Esposti (✉)
Dipartimento di Matematica, Università di Bologna, Piazza di Porta San Donato 5, I-40122 Bologna, Italy
e-mail: mirko.degliesposti@unibo.it

A. Celletti, U. Locatelli, T. Ruggeri and E. Strickland (eds.): *Mathematical Models and Methods for Planet Earth.* Springer INdAM Series 6, DOI 10.1007/978-3-319-02657-2_8,

along the text. In particular, I like to recall the attention to a very recent and relevant (in my opinion) application (by the same authors) to the intriguing case of the Voynich manuscript. Again, I do not pretend to be complete or original in the exposition, but I hope to capture some attention to the problem and to the cited references. Finally in Section 3, I briefly discuss some mathematical approaches to authorship attribution, recalling the problem and focusing on a very specific and recent case study I faced (in collaboration with D. Benedetto and G. Maspero) and involving Basil of Caesarea and his brother Gregory of Nissa, two influential 4th century Christian theologians, and the attribution of specific and discussed works in their corpora. This last section is heavily based on our recent paper [7] presented at the Workshop.

2 A new look at the old Zipf's law

Let us look at a given text x as a sequence of N words $x_1, x_2, \ldots, x_N$ over a vocabulary of d distinct words $D = \{\omega_1, \omega_2, \ldots, \omega_d\}$. For simplicity we assume that punctuation marks and other non alphabetic symbols have been removed, that words are any sequence of characters between two spaces and that they are kept in their original form. Of course other choices are possible, depending on the aims and on the applications: for example punctuation marks might play a role for authorship attribution, or words can be *stemmed* (a language dependent procedure) for some semantic extraction or topic classification tasks. The most elementary statistical property is clearly the frequency of different words in the text:

$$p_j = p(\omega_j) = \frac{1}{N}\sharp\{k : x_k = \omega_j\}$$

It is clear that this empirical frequencies, often tacitly confused with probabilities, are highly dependent on the text $p_j = p_j(x, N)$ and are obviously invariant by word shuffling. Together with the frequency, we also have a *rank* function $r : D \to \mathbb{N}$ that orders words with respect to their occurrence: $r(\omega) = 1$ for the most frequent word, $r(\omega) = 2$ for the second most frequent word, and so on.

It was already in 1949 that the philologist George Zipf in *Human behaviour and the principle of least effort*, after an extensive empirical study, found heuristic relations between the frequency and the rank: $p(\omega) = C \cdot r(\omega)^{-z}$, where usually z is slightly greater than 1 and C is a normalisation constant. In other words, in a double logarithmic scale the frequencies decays linearly as the rank increases. Being this the usual formulation of the famous *Zipf's law*, it is in fact equivalent to a previous observation made again by Zipf in *The Psycho-Biology of Language* (1936) that states that the number of words $N(n)$ which occur exactly n times decays as: $N(n) \sim n^{-\xi}$. Noting that for a word ω that appears exactly n times:

$$r(\omega) = \sum_{k=n}^{\infty} N(k) \sim \int_n^{+\infty} N(x)\, dx,$$

it is immediate to see that

$$z = \frac{1}{\xi - 1},$$

where in particular $z = 1$ for $\xi = 2$.

Besides the possible interpretations of this empirical law, it is worth to mention that as it is stated before, the scaling law between frequency and rank holds only on specific regimes that strongly depend on the corpus. As discussed by [34], the original form of the Zipf's law report above describe correctly the statistical behaviour of word frequencies only in the range middle-low to low of the rank variable: for a single long literary text this means a small window around $r \sim 100$ and $r \sim 2000$ which represent only a tiny portion of any significant literary vocabulary. The *non-Zipfian* behaviour of high rank (low frequencies) words observed as a robust feature in large corpora is not explained by the known model of the Zipf's law but it is conceivable related to non trivial aspects of the structural complexity of human writing. A more complete experiment performed by Montemurro in [34] on a large corpus of 2606 English group reveals the robustness of the non-Zipfian regime in the region of low frequent words (usually the ones related to the semantic content): while the *Zipfian region* extends (only) up to $r \sim 6000$ for this large corpus, while a second power regime, decreasing faster than the previous, appears clearly for high ranks words. Because of the poor statistics in the low frequency regime, it is not so easy to give a precise estimate of this second power low, even if a decay exponent around -2.3 as predicted by Montemurro is also quite compatible with the empirical observations by Ferrer and Solé on an english corpus formed by a large number of short texts (written and spoken) [18]. Since its introduction in linguistic, the Zip's law or in general power laws have been reproduced and investigate in physics, biology, earth and planetary sciences, economics and finance, computer science, demography and the social sciences, just to mention a few, essentially splitting in two the scientific community (very roughly speaking): the ones that believe that Zipf's law reveals fundamental functional properties of the systems on one hand, and the ones, exactly because of its broad universality, that argue against its relevance. There have been in fact many arguments against the meaningfulness or relevance of Zip's law [10, 12, 28, 33, 42].

On one hand, not only Zipf's law holds also for any shuffling of a real text, when any semantic or syntactic structure has been destroyed, but it also (at first sight) reproduced by almost trivial stochastic processes: consider for example any given finite alphabet of size $d \geq 2$ and add to it a character that corresponds to the *space*. Now construct a Bernoulli process (eventually not allowing two consecutive spaces) where each character and the space are independently generated with probability p and q respectively $(dp + q = 1)$. If we define *words* as any sequence of characters between two spaces, a good agreement with Zipf's law seems to hold, at least asymptotically in the length [28, 29]. But it was only recently that rigorous statistical tests on several types of random texts proved that the word rank histograms of these random tests are in fact inconsistent with those of real texts [20].

On the other hand, going back texts, one can interpret the Zipf's law following a qualitative explanation given by Zipf himself based on the *principle of least effort* that leads to an equilibrium between the *work/effort* of the speaker and the hearer, the two main actors of any communication process. This explanation enforces the re-use of already used words and leads to a model based on multiplicative stochastic process introduced by the sociologist Herbert Simon, few decades after Zipf's work (see [44] and also [19] and [13] for a more modern approach). However Simon's model is not able to reproduce power law exponent larger than one found in textual corpora of different languages and it is not capable to describe the *non Zipfian* behaviour of high rank words. In this model new words are created at a constant rate λ and, as noted in [45], it leads to a linear vocabulary growth $N(t) \sim \lambda t$, contrary to the empirical observation that indicate a sublinear growth $N(t) \sim \lambda t^\gamma$, with $0 < \gamma < 1$, where t denotes the total number of words and $N(t)$ is the total number of distinct words (*Heaps' law*, [27]). It is exactly in [45] that in order to reproduce realistic Zipf distributions, a new stochastic dynamical model has been introduced based on a non linear rate of generation of new words. This is an interesting model where the ratio of vocabulary growth is affected by two different processes that act at different scales: a global interplay between multiplicative and additive process in word selection that mimic small differences in vocabulary growth between texts of the same languages, depending on the context and on the author's style, and a more robust and local process that resemble grammar and syntactic complexity that in highly inflected languages (e.g. Latin) leads to faster proliferation of new words (in these languages several distinct words are produced by a common root just by declination and conjugation).

This is of course a sketchy and incomplete review of this huge subject. Personally I'm not even so sure I'm able to capture all the relevance of the discussions concerning the origin and nature of the Zipf's law, but certainly it stimulates the investigation of other statistical features of words (such as returning times or long correlations) and of other natural components of texts, such as the characters *n-grams*. This is certainly necessary if one is interested in one of the applications mentioned in the abstract. We perform a small step in this direction in the following section.

3 Entropy of words and semantic content: the case of the Voynich manuscript

As we have seen, Zipf's distribution captures the scaling law among frequencies of different words in a given text and it probably reflects global structural properties of languages. These global structures induce on one hand local short correlations between words, mostly due to grammar and syntactic rules, but on the other literary texts presents also long range correlations, going often beyond the size of thousand of words, reflecting the semantic content of the text. To bring an example deeply discussed in [2], take your favorite novel and consider the binary sequence obtained by mapping each vowel into a 1 and all other symbols into a 0. One can easily detect either structures on neighbouring bits or some repetition patterns on the size of words. But one should certainly be surprised and intrigued when discovering that there are structures (or memory) after several pages or even on arbitrary large scales of this binary sequence. In the last twenty years, similar observations of long-range correlations in texts have been related to large scales characteristics of the novels such as the story being told, the style of the book, the author, and the language [1,3,4,14,15,31,35,39], even if a complete explanation of these connections is still missing.

Recently in [2] we explain how long-range correlations flow from highly structured linguistic level down to the building blocks of a text (words, letters, etc...), but while we refer the interested reader to the paper for more mathematical details and experiments (*War and Peace* by L. Tolstoy is shown in the paper, while other international novels are explored in the Supporting Information (SI)), here we like to focus our attention to a very recent and nice paper by Montemurro and Zanette that approaches a concrete case (the *Voynich manuscript*) with the use of simple but powerful ideas regarding returning time distribution of words. Again, the presentation will be sketchy and incomplete, but I hope it can capture curiosity towards the use of quantitative methods for approaching concrete philological problems. In the last section I will discuss another concrete example in authorship attribution. A nice theoretical approach with relevant practical applications to automatic keywords extraction has been introduced already some time ago in [35] where the attention has been shift from the simple frequency to the *returning time distribution of a given word*, namely the integer sequence given by the number of words one has to read before meeting the given word again. In this paper, starting from Shannon's mutual informations, the authors proposed a measure that captures the relationship between the statistical structure of words sequences and their semantic contents. It is evident that different words appear with different frequencies in different part of the texts and we now know that the heterogeneity in the distribution of individual words is related to the linguistic role of the world and also to the specificity of that word to certain sections or parts of the text: while a *functional* word (i.e. a word with little lexical meaning, such as prepositions, pronouns, auxiliary verbs, conjunctions, grammatical articles or particles) almost evenly distributes among the different section of a text, *content* words on the opposite tend to *clusterize* on specific parts of the texts. Namely, the actual *spatial* distribution of content words with specific semantic meaning among a given text *deviates* considerably from the typical spatial distribution that the same word would have if randomly shuffling all the words in the text. To be more precise, following again [35], consider again a text x as a sequence of N words $x_1, x_2, \ldots, x_N$ over a vocabulary of d distinct words $D = \{\omega_1, \omega_2, \ldots, \omega_d\}$ and assume that the text is divided into P equal parts of (word) length $s = N/P$. Given a word ω, let $n_j = n_j(\omega)$ be the number of occurrences of word ω in the j-th part of the text ($\sum_{j=1}^{P} n_j(\omega) = n(\omega)$, the total occurrence of the word in the text): $p(\omega|j) := n_j/s$ can be interpreted as the probability of finding the word ω in part j, $j = 1, \ldots, P$ (note that $\sum_{k=1}^{d} p(\omega_k|j) = 1$) and we can define two natural quantities:

- An *entropy of the word* associated to its distribution with respect to the given partition of the text ($n = n(\omega)$)

$$H_P(\omega) := -\sum_{j=1}^{P} \frac{n_j}{n} \log_2 \frac{n_j}{n}.$$

- An information measure associated to the word

$$\Delta I_P(\omega) := p(\omega)\left[\langle \hat{H}_P(\omega)\rangle - H_P(\omega)\right],$$

where $p(\omega) = n/N$ is the overall frequency and $\langle \hat{H}_P(\omega) \rangle$ is the same entropy computed on a randomly shuffled version of the text and averaged over all possible realisation of the shuffling (computed analytically in [35]).

If words are ranked with respect to the information measure, the top words corresponds to those more closely related to the semantic contents and can be considered as *keywords*. Several experiments confirm the practical validity of this approach, even if we believe that a more systematic evaluation procedures must be performed in order to compare this method for keywords extraction with respect to other traditional (i.e. more based either on linguistic or computational linguistic methods) extraction methods. Some personal experiments (not published) also indicate an interesting cross-language stability of the method, yielding equivalent results when keywords extraction is applied to the same text written in different languages. While the parameter P that defines the partition text can be introduced a priori by considering natural scales of the text (e.g.: chapter, sections or paragraphs), it can also be automatically estimated by a variational procedure that maximise the overall information content of the text:

$$\Delta I(P) := \sum_{\omega=1}^{d} \Delta I_P(\omega) = \sum_{\omega=1}^{d} p(\omega) \left[\langle \hat{H}_P(\omega) \rangle - H_P(\omega) \right].$$

The previous method recently finds a very interesting application in a nice paper by Montemurro and Zanette where keywords have been extracted in the particular and intriguing case of the Voynich manuscript [36]. This is a very interesting case that poses fascinating challenges to everyone interested in mathematical and statistical properties of languages. It is in fact a manuscript of about 250 pages, divided in 5 sections (Herbal, Astrological, Biological, Pharmacological and Recipes), with a clear medieval-codex look (carbon dating bring back its origin to the second half of the fifteenth century) that was owned by the Polish-American antiquarian Wilfrid Voynich since 1912 until his death in 1930. As described in [36], today the manuscript belongs to the Beinecke Rare Book and Manuscript Library of Yale University and can be browsed online through the Wikimedia commons at the address *commons.wikimedia.org/wiki/Voynich_manuscript*. It contains nice illustrations of unknown plants, astronomical or astrological diagrams, and naked nymphs, bathing in strange arrangements of pools or tubs connected by complex systems of pipes. But what makes the manuscript so intriguing is the fact that not only the author, but also the language used in the text are unknown and resisted any attempt of translation or decryption: it appears to be written in an alphabet of about 40 symbols arranged into *words* of variable length and it has been converted to a digital file using the so called European Voynich Alphabet (EVA) [16]. Basically three (overlapping) hypothesis can be made about the manuscript [40]:

- *Cipher text hypothesis*: the manuscript contains natural language text (for example Latin or German) that has been non trivially encrypted.
- *Plain text hypothesis*: the manuscript is plain text in natural, not yet identified language that either did not possess an original alphabet in the beginning 16th

century or the system of writing appeared too complex to a medieval scholar. The word length statistics makes East Asian languages, in particular Chinese, the most promising candidate for this (Chinese theory). Alternatively, the script could also have been invented together with an artificial language.
- *Hoax hypothesis*: the manuscript contains no meaningful text at all and it was a fabrication during Renaissance time to attract rich book collectors, quite common at that time.

Most of the quantitative investigations of the manuscript are based on a character scale, where the manuscript is seen as a sequence of characters (space included) and several statistical analysis tools have been used: from random walk techniques to autocorrelation measures and higher order character entropy (see for example [40] and [23]). But it was exactly in [36] that essentially with the use of the entropy of words previously discussed the Voynich manuscript have been investigated by addressing the large scale organisational structure which results from the distribution of words over the whole text. Referring the interested reader to [36] for further details, it is interesting to notice that this new approach allows to extract the *most informative words*, even if we do not know their meaning and to support at the end the hypothesis that the manuscript possesses a structured information content.

4 Authorship Attribution: a case study

I believe that it is now natural to wonder if we can push investigations a little bit further. For example, can we use quantitative methods to recognise the author of a given text, at least when we can restrict to a finite number of possible authors? This is essentially the aim of Authorship Attribution (A.A.) [22].

Research in A.A. seems to be a perfect meeting point for different sciences: information theory, statistical computations, mathematical models and more in general quantitative analysis have an object of study that is typical also of very different disciplines, like literature studies, philology, history and in general humanities [41]. The interaction between philological and mathematical sciences in the analysis of the problem is not only at the level of definition of the case and of discussion of the results, but at the very level of development of the methods and of the experimental setting. For this interaction to be fruitful it is then mandatory to attack concrete problems and not artificial ones. This could prelude to a loss of generality in the research objectives and might appear a very narrow strategy. On the contrary, I'm convinced that probing methods and techniques on very specific problems might open new insight into fundamental and general results. This is exactly the case for the very specific two authors problem studied in [7]: Basil of Caesarea [38] and his brother Gregory of Nyssa [32] were two important bishops and theologians of the 4th century. Their thoughts and teachings were fundamental for the definition of Christian doctrine of the Trinity, i.e. of God as one divine nature and three divine persons.

Their discussion was almost thirty years long and many works were devoted to it. *Epistula 38* (Ep. 38) is one of them. It is a letter that was transmitted in *Basil's epistolary corpus* [17], but that has been attributed also to his brother Gregory of Nyssa as a dogmatic treatise addressed to their brother Peter (see [7] for more details on this dispute).

The Corpus used in the study is composed by all the known works by Basil and Gregory of Nyssa, The digitalized texts in *Thesaurus Linguae Graecae* (TLG) have been used [47] and a unique coding and a common cleaning procedure has been defined to normalise the texts. We refeer to [7] for more details on the cleaning procedure and also for a detailed description of the corpus (about 10 millions characters) and its use. A first crucial assumption beyond all of our mathematical techniques is that a text should be thought as a sequence of symbols chosen from an alphabet, while the author is interpreted as a *source* of literary texts. Assuming that the text is "just" a symbol sequence means not taking into consideration either the semantic content of the text or its linguistic/syntactic/grammatical aspects: letters of the alphabet, punctuation marks, blank spaces between are just abstract symbols, without any hierarchy. The two methods we have implemented here are in fact the development of two methods that have been already used in a project concerning the attribution of Antonio Gramsci's papers [5]. Essentially each method defines a kind of *similarity distance* between texts: given any pair of texts, each method produces a positive number we interpret as the distance between the texts.

Let us very briefly describe both algorithms, referring to [7] for further details. The first method we used based on (characters) n-grams is probably one of the simplest possible measures on a text, and it has a relatively short history in published bibliography: after a first experiment based on bigram frequencies presented in 1976 by W.R. Bennett [9], V. Kešelj et al [21] published in 2003 a paper in which n-grams frequencies were used to define a similarity distance between texts (see also [11]). Here we present the version of the similarity distance as introduced and discussed in [5]: we call ω an arbitrary n-gram, and we denote by $f_X(\omega)$ ($f_Y(\omega)$) the relative frequency with which ω occurs in text X (Y). $D_n(X)$ is the *n-gram dictionary* of the text X, that is, the set of all n-grams which have non-zero frequency in X (similarly for Y) and we define what we will call the n-gram distance between text X and text Y as[1]:

$$d_n(X,Y) := \frac{1}{|D_n(X)| + |D_n(Y)|} \sum_{\omega \in D_n(X) \cup D_n(Y)} \left(\frac{f_X(\omega) - f_Y(\omega)}{f_X(\omega) + f_Y(\omega)} \right)^2. \quad (1)$$

Here, $|D_n(X)|$ and $|D_n(Y)|$ are the numbers of different n-grams in the two dictionary, respectively, and the sum is taken over all the different n-grams occuring in the two texts.

The second method we use to estimate the similarity between texts is based on *data compression* and its role in the estimation of the entropy of a source. Data

[1] To be more precise, d_n is a pseudo-distance, since it does not satisfy the triangular inequality and it is not even positive definite: two texts X,Y can be at distance $d_n(X,Y) = 0$ without being the same.

compression is nowadays a very well established field of information theory, thanks to the founding papers published by J. Ziv, A. Lempel and their coworkers in the 1970s (cf., among others, [24–26] and the review paper [43]). They proposed a variety of compression algorithms (the family of) *LZ algorithms*, based on the idea of a clever *parsing* (subdivision) of the symbolic sequence, i.e. to split it up into pieces so that this separation can then be used to produce a shorter, equivalent version of the string itself. In 1993 J. Ziv and N. Merhav [46] proposed a method to estimate the relative entropy (or Kullback-Leibler divergence) between couples of information sources. The relative entropy is basically a measure of the similarity between the information emitted by the sources and they proved that a modified version of an LZ algorithm, where the subsequences for a sequence are searched in another sequence, can be used to approximate the relative entropy between the two sources that generated such sequences. This important result was used in various subsequent studies, among which [8] and [5], to deal with problems of text classification and clustering. Following these ideas, we have introduced a new algorithm to estimate the similarity between two symbolic sequences over a common alphabet (we refer to the main paper for details about its implementation). This leads to a true compressor program which effectively shows better compression rates with respect to `gzip`, as shown in Table 1 in [7]. The details of this method are designed to optimize the compression ratio in order to obtain a good estimate of the relative entropy between texts. Because of this specific optimization, the complexity of the implementation naturally increases.

A key point is that our methods are both quite sensitive to the length of the reference texts (in particular the entropic one). We faced this serious problem by using a cutting procedure, together with (in few words) a new probabilistic ranking method for attribution, in order to transform a problem of managing texts of different sizes, in the problem of managing a different number of texts of the same size. In fact, our procedure has been designed to work on two different scales of text length (again, see [7] for details). Using the previous two methods (n-grams and entropy) on a large *controlled corpus* of more than 200 works with known attribution, we were then able to test the efficiency of the procedure. We just report here the following table from [7] that shows the final results in terms of standard indexes from information retrieval we have been used to measure the efficiency of our method:

	Basil	*Gregor*
recall	0.87	0.90
precision	0.96	0.93
F-measure	0.91	0.91

Concerning Ep. 38, it was not included in the previous computations, as many other works, because its authorships is discussed. But after having verified the efficiency of the method in discriminating between the two authors, a careful investigation of the disputed works has been performed. In particular Ep. 38 has been strongly attributed to Gregory and the quantitative results have been interpreted also through philological considerations that we omit here, together with the other details. This result seems to suggest that the combination of philological and numerical techniques in the design of the method is very effective. As far as we know, our method is the first one able to analyse a real philological problem in authorship attribution of ancient Greek works and to give a clear result, in agreement with what was previously known in scientific literature [30].

References

1. Allegrini, P., Grigolini, P., Palatella L.: Intermittency and scale-free networks: a dynamical model for human language complexity. Chaos, Solitons and Fractals **20**, 95–105 (2004)
2. Altmann, E.G., Cristadoro, G., Degli Esposti, M.: On the origin of long-range correlations in texts. Proceedings of the National Academy of Sciences **109**, 11582–11587 (2012)
3. Alvarez-Lacalle, E., Dorow, B., Eckmann, J.P., Moses E.: Hierarchical structures induce long-range dynamical correlations in written texts. Proc Natl Acad Sci USA **103**, 7956-7961 (2006)
4. Amit, M., Shmerler Y., Eisenberg, E., Abraham, M., Shnerb N.: Language and codification dependence of long-range correlations in texts. Fractals **2**, 7-13 (1994)
5. Basile, C., Benedetto, D., Caglioti, E., Degli Esposti M.: An example of mathematical authorship attribution. J. Math. Phys. **41**, 125–211 (2008)
6. Benedetto, D., Caglioti, E., Degli Esposti, M.: The unreasonable effectiveness of Mathematics in Human Science: the attribution of texts by Antonio Gramsci. In: Emmer, M.(ed.) Imagine Math - Between Culture and Mathematics, pp. 143–154. Springer-Verlag Italia, Milano (2012)
7. Benedetto, D., Degli Esposti, M., Maspero, G.: The puzzle of Basil's Epistula 38: a mathematical approach to a philological problem. Journal of Quantitative Linguistics (2013)
8. Benedetto, D., Caglioti, E., Loreto, V.: Language trees and zipping. Physical Review Letters **88**(4), 48702 (2002)
9. Bennet, W. R., Scientific and engineering problem-solving with the computer. Prentice-Hall, Englewood Cliffs, NJ (1976)
10. Bernhardsson, S., da Rocha, L.E.C., Minnhagen, P.: The meta book and size-dependent properties of written language. New Journal of Physics **11**, 123015 (2009)
11. Clement, R., Sharp, D.: Ngram and Bayesian classification of documents for topic and Authorship. Lit. Ling. Comp. **18**, 423 (2003)
12. Conrad, B., Mitzenmacher, M.: Power laws for monkeys typing randomly: the case of unequal probabilities. IEEE Transactions on Information Theory **50**, 1403–1414 (2004)
13. Dickman, R., Moloney, N.R. , Altmann, E. G.: Analysis of an information-theoretic model for communication. Journal of Statistical Mechanics: Theory and Experiment **12**, 12022 (2012)
14. Ebeling, W., Neiman, A.: Long-range correlations between letters and sentences in texts. Physica A **215**, 233–241 (1995)
15. Ebeling, W., Pöschel, T.: Entropy and long-range correlations in literary English. Europhys Lett **26**, 241–246 (1994)
16. http://www.voynich.nu/extra/eva.html
17. Fedwick, P.J.: Bibliotheca Basiliana Universalis I. Brepols-Turnhout, pp. 620–623, 674–678 (1993)
18. Ferrer i Cancho, R. , Solé, R.V.: Two regimes in the frequency of words and the origins of complex lexicons: Zipf's law revisited. Journal of Quantitative Linguistic **8**(3), 165–173 (2001)
19. Ferrer i Cancho, R. , Solé, R. V.: Least effort and the origins of scaling in human language. Proc. Natl. Acad. Sci. **100**(3), pp. 788–791. USA (2003)
20. Ferrer i Cancho, R., Elvevag, B.: Random texts do not exhibit the real Zipf's law-like rank distribution. PLoS ONE **5**, 9411 (2010)

21. Kešelj, V., Peng, F., Cercone, N., Thomas, C.: N-gram-based author profiles for authorship attribution. In: Kešelj, V., Endo, T. (eds.) PACLING'03. Proceedings of the Conference Pacific Association for Computational Linguistics, pp. 255-264. Dalhousie University, Halifax (2003)
22. Juola, P.: Authorship Attribution. FNT in Information Retrieval **1**, 233–334 (2007)
23. Landini, G.: Evidence of linguistic structure in the Voynich manuscript using spectral analysis. Cryptologia **25**, 275–295 (2001)
24. Lempel, A., Ziv, J.: On the complexity of finite sequences. IEEE Trans. Inform. Theory IT **22**(1), 75-81 (1976)
25. Lempel, A., Ziv, J.: A universal algorithm for sequential data compression IEEE Transactions on Information Theory **23**(3), 337–343 (1977)
26. Lempel, A., Ziv, J.: Compression of individual sequences via variable-rate coding. IEEE Transactions on Information Theory IT **24**(5), 530–536 (1978)
27. Lü, L., Zhang, Z.K., Zhou, T.: Zipf's law leads to Heaps' law: analyzing their relation in finite-size systems. PLoS ONE **5**, e14139 (2010)
28. Li, W.: Random texts exhibit Zipf's-law-like word frequency distribution. IEEE T Inform Theory **38**, 1842–1845 (1992)
29. Mandelbrot, B.: An informational theory of the statistical structure of language. In: Jackson, W. (ed.) Communication Theory. Butterworths, London (1953)
30. Maspero, G., Leal, J.: Revisiting Tertullian's Authorship of the Passio Perpetuae through Quantitative Analysis. In Grzybek, P. (ed.) Text and Language. In: Kelih, E., Maoutek, J. (eds.) Structures – Functions – Interrelations Π Quantitative Perspectives, pp. 99–108. Wien (2010)
31. Melnyk, S.S., Usatenko, O.V., Yampolskii, V.A.: Competition between two kinds of correlations in literary texts. Phys Rev E **72**, 026140 (2005)
32. Meredith, A.: Gregory of Nyssa. Routledge, London, New York (1999)
33. Mitzenmacher, M.: A brief history of generative models for power law and lognormal distributions. Internet Mathematics **1**, 226–251 (2003)
34. Montemurro, M.A.: Beyond the Zipf–Mandelbrot law in quantitative linguistics. Physica A: Statistical Mechanics and its Applications **300**, 567–578 (2001)
35. Montemurro, M.A., Zanette, D.: Towards the quantification of the semantic information encoded in written language. Adv Comp Syst **13**, 135–153 (2010)
36. Montemurro, M.A., Zanette, D.: Keywords and Co-Occurrence Patterns in the Voynich Manuscript: An Information-Theoretic Analysis. PLoS ONE **8**, e66344 (2013)
37. Newman, M.E.J.: Power laws, Pareto distributions and Zipf's law. Contemporary physics **46**, 323–351 (2005)
38. Rousseau, P.: Basil of Caesarea. University of California Press, Berkeley (CA), Los Angeles (CA), London (1998)
39. Schenkel, A., Zhang, J., Zhang, Y.: Long range correlation in human writings. Fractals **1**, 47–55 (1993)
40. Schinner, A.: The Voynich Manuscript: Evidence of the Hoax Hypothesis. Cryptologia **31**, 95–107 (2007)
41. Stamatatos, E.: A Survey of Modern Authorship Attribution Methods. Jour. Am. Soc. Infor. Sci. Tech. **60**, 538–556 (2009)
42. Suzuki, R., Tyack, P.L., Buck, J.: The use of Zipf's law in animal communication analysis. Anim. Behav. **69**, 9–17 (2005)
43. Wyner, A.D., Ziv, J., Wyner, A.J.: On the role of pattern matching in information theory. IEEE Transactions on information Theory **44**(6), 2045–2056 (1998)
44. Zanette, D.: Statistical Patterns in Written Language, available at http://fisica.cab.cnea.gov.ar/estadistica/zanette/
45. Zanette, D., Montemurro, M. A.: Dynamics of text generation with realistic Zipf's distribution. J Quantitative Linguistics **12**, 29–40 (2005)

46. Ziv, J., Merhav, N.: A measure of relative entropy between individual sequences with application to universal classification. IEEE Transactions on Information Theory **39**(4), 1270–1279 (1993)
47. The digital library has been developed by University of California, Irvine. http://www.tlg.uci.edu/

Space Debris Long Term Dynamics

Anne Lemaitre and Charles Hubaux

Abstract Our lifestyle is strongly dependent on the presence of spacecraft: telecommunications, GPS or cellular phones, TV, Internet, climate watches, ecological studies, catastrophe prevention, military surveys, ... Despite the technological progress, the costs and the risks due to the space debris are increasing and can really stop or drastically reduce the systematic replacement or extension of the present satellite constellations, stopping the worldwide communication.

In the next years, a special attention should be dedicated to the space debris problematic, to protect the space environment and to allow technological innovations, in the present framework of sustainable development.

In particular, more precise information about the dynamics and the behavior of debris has to be collected; the methods and theories of classical celestial mechanics are very suitable to describe the long term dynamics of these debris. Interesting results have been obtained by new approaches of the problem: the resonant description of the dynamics of geosynchronous debris, the consideration of the solar radiation pressure as an important perturbation, specially for objects with a large A/m coefficients and, because of the lifetimes of the debris, the use of symplectic integrators, as for the natural bodies but adapted to the specific force model.

1 Introduction

The term *space debris* is used to characterize all non functional objects (fragments of satellites, rocket parts, remains of explosions or collisions), of all sizes and all chemical compositions, which orbit around the Earth at different altitudes.

A. Lemaitre (✉)
naXys, University of Namur, Rempart de la Vierge 8, B 5000 Namur, Belgium
e-mail: anne.lemaitre@unamur.be

C. Hubaux
naXys, University of Namur, Rempart de la Vierge 8, B 5000 Namur, Belgium

A. Celletti, U. Locatelli, T. Ruggeri and E. Strickland (eds.): *Mathematical Models and Methods for Planet Earth.* Springer INdAM Series 6, DOI 10.1007/978-3-319-02657-2_9,

The number of space debris has dramatically increased in the last decades; since Sputnik in October 1957, more than 6 000 satellites have been launched and about 200 exploded in space, either accidentally or on purpose. Indeed for strategic or political reasons, especially during the Cold War, the nations did not hesitate to provoke the explosion of a satellite to protect their innovative technological secrets. Moreover space missions did not care about the situation of satellites after their lifetime and some of them are orbiting the Earth for years after their official inactivity date. Only a small fraction of the objects in orbits are still active satellites ...

The drag forces will rapidly bring back to the Earth most of the lowest objects, characterized by altitudes between 500 and 900 kilometer (LEO, Low Earth Orbits). Reminding that the lifetime of a classical satellite is estimated to a month for an altitude of 300 km, a year for 400 km, 10 years for 500 km, decades for 700 km, centuries for 900 km and millennia for 1200 km, the process is not fast. Even if the drag is able to clean progressively the LEO region, the presence of a huge number of debris is responsible for collisions and consequently, generates a continuous re-population of the region. For the highest ones, in particular for the very popular geosynchronous orbits (GEO) situated at 36 000 kilometer of altitude and characterized by periods of exactly 24 hours, the drag is inefficient, the objects orbit for centuries, and the size of the space is not anymore a convincing argument.

Let us visualize the situation in two pictures: Figure 1 draws the number of debris with respect to their distance to the Earth surface (altitude), with the three peaks

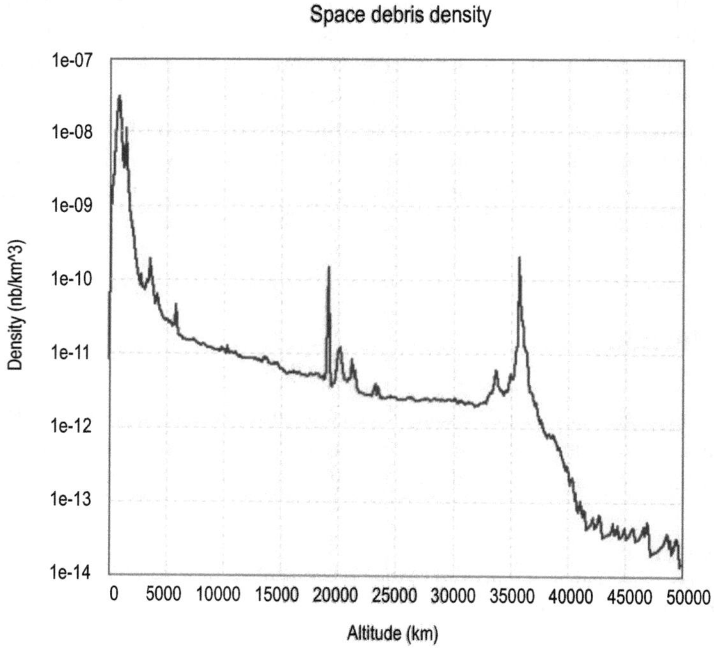

Fig. 1 Density of Earth orbiting space objects with respect to their altitude. From [15]

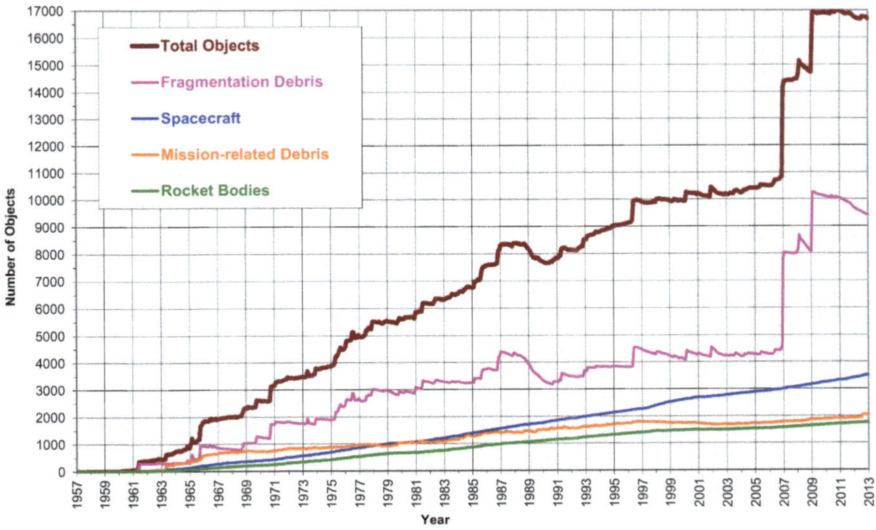

Fig. 2 Growth of debris (in the TLE catalogue) from 1957 to 2012. From NASA, www.nasa.gov

of concentration, for the LEOs, for the GEOs and, in between, for the very popular 12 hours period orbits (not far from the constellations GPS or GALILEO). Figure 2 represents the growth of listed objects with respect to time, from 1957 to nowadays. Two recent catastrophic events are clearly visible on the graph: the deliberate explosion of a Chinese satellite in 2007 (Fengyun)[1] and the accidental collision between two satellites, the active American telecommunication Iridium 33 and the inactive military Russian satellite Cosmos 2251 in 2009[2].

These two figures only concern the listed objects; indeed a catalogue of about 16 000 debris (20 000 in 2014) is maintained and completed by NASA at each registered collision. It contains the objects larger than 10 centimeter for LEOs (and larger than 1 meter for GEOs) which could really damage an active satellite and compromise its mission. They are stored in a file called *the two line elements* or *TLE* as reference to their format. More precisely, a two-line element set is a data format used to convey sets of orbital elements that describe the orbits of Earth-orbiting objects. A computer program called a *model* can use the TLE to compute the position of any satellite or debris at any particular time.

---

[1] The 2007 Chinese anti-satellite missile test was conducted by China on January 11, 2007. A Chinese weather satellite, the FY-1C polar orbit satellite of the Fengyun series, at an altitude of 865 kilometer with a mass of 750 kg was destroyed by a kinetic kill vehicle traveling with a speed of 8 km/s in the opposite direction.

[2] The collision occurred on February 10, 2009, at 789 kilometer above the Taymyr Peninsula in Siberia, when Iridium 33 and Kosmos 2251 collided at a speed of 11.7 kilometer per second. The Iridium satellite was an American telecommunication satellite member of a constellation, it was still operational at the time of the collision, while the Russian military satellite had been out of service since at least 1995 and was no longer actively controlled.

The risk of collision is real and avoidance maneuvers are performed regularly, by the active satellites, the space shuttles or the ISS, increasing the cost of the missions by consuming fuel.

However the TLEs only refer to the huge objects and represent the tip of the iceberg. Small-size debris (resulting primarily from fragmentations or slag and dust residues from solid rocket motor firings) are much more difficult to detect. Various approaches described in [15] have been used to assess their number. The estimated number of debris is around 670 000 for sizes larger than 1 cm and more than 170 million for sizes larger than 1 mm. These debris are neither catalogued nor individually identified and even if we stopped all launches, the number of debris would still increase for several years, just by collisions and fragmentations of the present objects in orbits.

Special equipments and armor plating protections are now systematically scheduled for the spacecrafts, increasing their cost and requiring ever more powerful rockets (because of their weight). Even if they are expensive, these protections are efficient for the small debris, below the centimeter size, but not at all for larger ones. This is why, presently, the most dangerous population is the intermediary one, between 1 and 10 centimeter, where the objects are too small to be followed individually but are too big to be considered only as a dusty environment altering the surfaces.

2 Recent contributions of celestial mechanics

At the end of the nineties, the space debris population has drawn the attention of the celestial mechanics community, traditionally involved in the dynamics of natural bodies. Indeed active probes have lifetimes of a few months or years, require a high precision orbit and a full model, and are controlled by maneuvers to keep their efficiency; the problems to solve are then very specific, mainly numerical, and slightly different from natural bodies dynamics. At the opposite, space debris are present for centuries, uncontrolled, and migrate everywhere following the perturbations, similarly to natural bodies. Their stability, their global behavior, their long term evolution, can be described by means of the same tools as other populations, as minor planets, particles in rings, exoplanets, etc.

2.1 A resonant approach of the geosynchronous orbit

The geosynchronous orbit can be considered as a resonant situation (a complete revolution of the satellite is performed in exactly the same time as a rotation of the Earth, which means that the two corresponding frequencies are almost equal). By averaging over the fast angles, a long term *resonant* motion, with a period of about 818 days, can be deduced, as well as an amplitude of the libration zone (see for example [2] and [20]). Figure 3 gives a representation of the resonant motion in a rotating

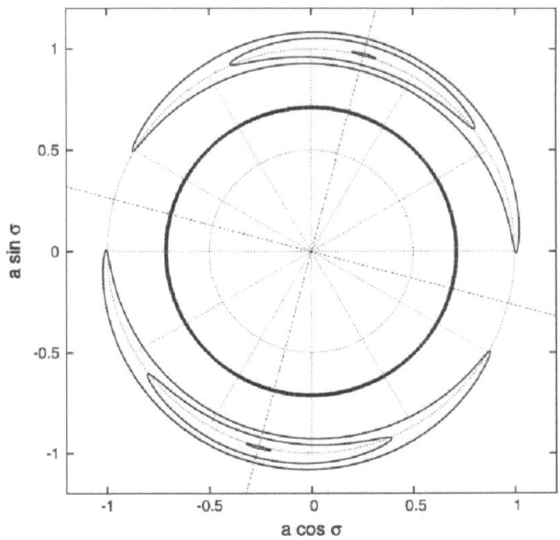

Fig. 3 Motion of geosynchronous space debris near stable equilibria in a rotating reference frame as seen from the pole. The variations on the mean semi-major axis have been amplified for the illustration by a factor 100, the unit being the exact geostationary radius. From [20]

frame (moving with the Earth) in the plane $(a\cos\sigma, a\sin\sigma)$, where a is the mean semi-major axis and σ the resonant angle, difference of the two synchronous angles. The width of the pendulum like resonant zone is about 70 km. This simple resonant model can be taken as a first averaged integrable model, on which perturbations can be added. The motion of geosynchronous debris can be described, up to a specific order, by a classical perturbation theory and the integration on long periods of time (several tens of years) is correctly performed, with a suitable elimination of the short periodic terms. A similar work has been performed for the 12 hours objects, blocked in a 1:2 resonance with the rotation of the Earth ([16]).

2.2 The large A/m objects

Observations [17] have shown an unexpected concentration of geosynchronous debris on highly eccentric orbits; they are characterized by very large values of A/m (area over mass), a property which increases the effect of the solar radiation pressure on their dynamics. This force can become the first perturbation after the two body problem for light and large debris. In this case, this force introduces large variations in their eccentricity (with a period of one year) and in their inclination (period of dozens of years). The motion of such a body can be described by simple formulae in eccentricity and inclination (first given in [4] and completed by [21]:

$$e\sin\varpi \simeq \kappa\sin\lambda_{\odot} - C_x \tag{1}$$

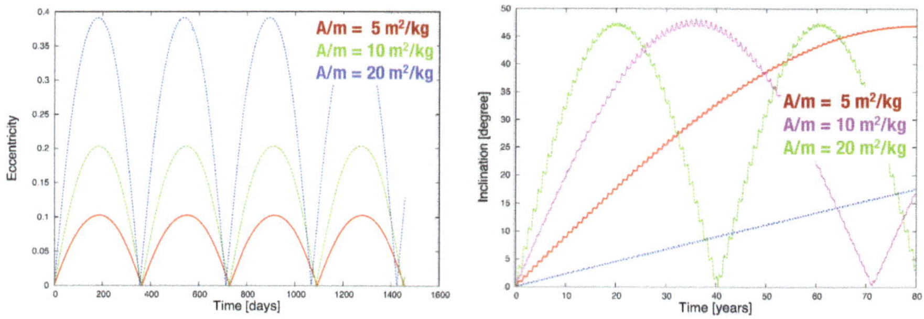

Fig. 4 Behavior of the eccentricity and of the inclination for very large values of A/m. From [19]

$$e \sin \varpi \simeq \kappa \cos \lambda_\odot \cos \varepsilon + C_y \tag{2}$$

$$\sin \frac{i}{2} \sin \Omega \simeq -E \sin \psi \tag{3}$$

$$\sin \frac{i}{2} \sin \Omega \simeq E \cos \psi + F \tag{4}$$

with ε the obliquity, $\lambda_\odot$ the longitude of the Sun (in a circular orbit), e the eccentricity, i the inclination, ϖ the longitude of the perigee and Ω the longitude of the ascending node, $\psi = \nu t + \psi_0$. The constants C_x, C_y, F and ψ_0 are determined by the initial conditions. κ is a constant proportional to $\frac{A}{m}$ and ν is proportional to its square, $(\frac{A}{m})^2$.

As shown in Figure 4, the eccentricity (starting from 0) can reach values up to 0.1 for $A/m = 5m^2/kg$, 0.2 for $A/m = 10m^2/kg$ and 0.4 for $A/m = 20m^2/kg$, while the inclination always reaches the same maximal value (close to twice the obliquity, about 46°) but for periods shorter and shorter as A/m increases : 200 years for $A/m = 5m^2/kg$, 70 years for $A/m = 10m^2/kg$, and 40 years for $A/m = 20m^2/kg$.

Moreover using simple perturbation formulæ, the motion of the eccentricity can be considered as a short periodic motion on the inclination, and can be superposed to the long term motion of the inclination as shown in [10]. On a simple model (two-body problem and solar radiation pressure), this approach has been compared to a complete numerical integration and gives an excellent approximation, as shown in Figure 5.

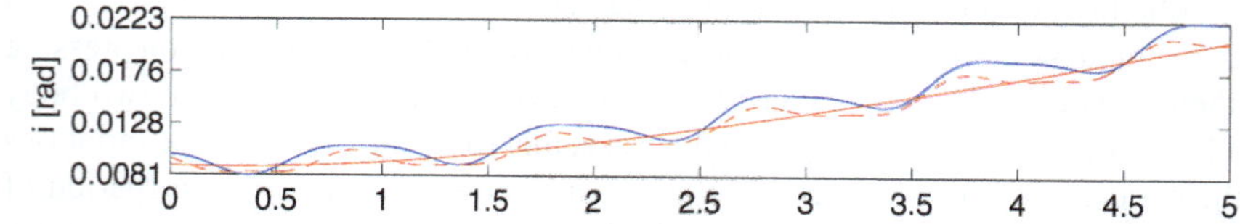

Fig. 5 Behavior of the inclination on five years: averaged motion (continuous line), perturbed averaged motion (dashed line) and numerical integration (curved continuous line). From [10]

2.3 Chaos indicator MEGNO and frequency map

Chaos indicators are suitable tools to determine the long term dynamics of geosynchronous space debris. In particular, we use the Mean Exponential Growth factor of Nearby Orbits or MEGNO [5], [6]. The first application of MEGNO to the space debris was performed by [3].The MEGNO is easy to implement, from the equations of motion combined with the variational equations, and gives an immediate answer about chaos: if the mean MEGNO equals 2, the orbit is stable, if it is higher, the orbit is chaotic [5]. The dynamics includes the central body attraction, the second degree and order harmonics (J_2, C_{22} and S_{22}), the luni-solar interactions as well as the solar radiation pressure (see [22] or [11]). Figure 6 shows an example of such a stability map, obtained for the geosynchronous region, where the pendulum-like shape is easily recognizable, signature of the 1:1 gravitational resonance.

These computations detected, for large A/m values, islands or curves corresponding to secondary resonances, defined as local commensurabilities between the resonant angle σ, with its period of 818 days, and the angle $\lambda_{\oplus}$ and its period of 365 days. Thanks to the Frequency Map Analysis FMA [13], applied for the first time in this context, more detailed maps were obtained and each secondary resonance could be identified, in the libration and circulation zones of the pendulum, as shown in Figure 7. The chaos indicator is the second derivative of the FMA, stable zones are colored in blue and the most chaotic appear in red. In particular the map revealed the presence (in the libration zone) of three stable islands inside the pendulum libration zone.

A complete mathematical analysis, based on a succession of canonical transformations, in suitable action-angle variables, confirms this numerical result (the positions, the widths and the periods) [14]. These regions are especially interesting

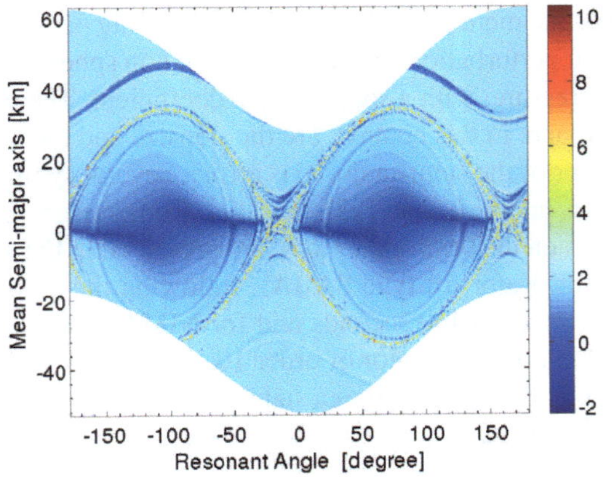

Fig. 6 The MEGNO computed as a function of initial mean longitudes λ_0 and semi-major axis a_0 of the debris, in the plane (σ, a). The mean longitude grid is $1°$ and the semi-major axis grid is 1 km spanning the 42164 ± 35 km range. The initial conditions (at 21 March 2000) are $e = 0.1$, $i = 0.004\circ$, $\Omega = \varpi = 0°$, with an area-to-mass ratio $A/m = 10m^2/kg$. From [22]

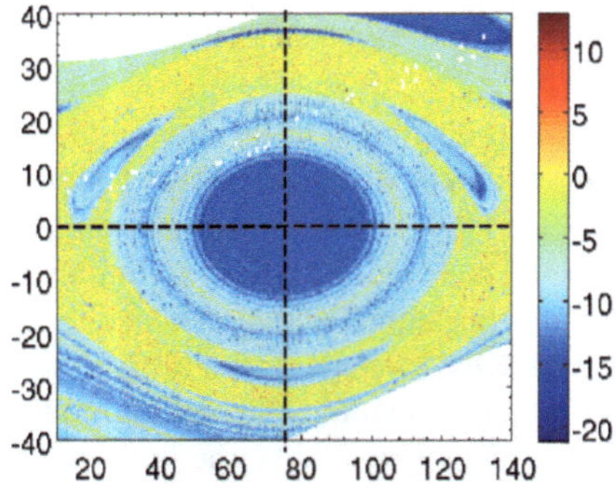

Fig. 7 Example of a chaos map obtained by the Frequency Map Analysis for the geosynchronous region, in the plane (σ, a) where σ is given in degrees and a in kilometer with respect to the exact geosynchronous position. The bluer the indicator, the stabler the orbit. From [14]

because the debris can accumulate and stay for centuries, trapped in these secondary resonances, ideal situations for *churchyard* or *parking* orbits for the old satellites and for future ADR (Active Debris Removal).

2.4 Suitable integrators

To reproduce the motion of space debris on very long periods of time, different techniques can be used: an averaged approach (MEAN) for long term dynamics of geosynchronous objects, taking the resonance into account in the averaging process [19], an osculating integrator adaptable to any telluric planet (NIMASTEP) and pushing the numerical integration to high orders [7] and a symplectic one (SYMPLEC) which avoids the increase of the energy on very long periods of time. They all include the luni-solar perturbations, the effects of the non sphericity of the Earth and the solar radiation pressure, with a specific A/m coefficient. The two advantages of the symplectic approach, in this context, are the stepsize and the quasi conservation of the energy error (or of the semi-major axis error). Several symplectic approaches have been compared, adapted from [23] and [12]. Comparisons between NIMASTEP and SYMPLEC, on the semi-major axis, are presented in Figure 8. The stepsize for the symplectic integrator (of type $SABA_4$, following the classification of [12]) was chosen as 4 hours or 14400 seconds and for NIMASTEP (associated with an Adams-Bashforth-Moulton integrator of order 10, ABM_{10}) as 1152, 1004, 864 and 432 seconds. Only this last choice gives a stability of the semi-major axis, after 200 years, similar to SYMPLEC.

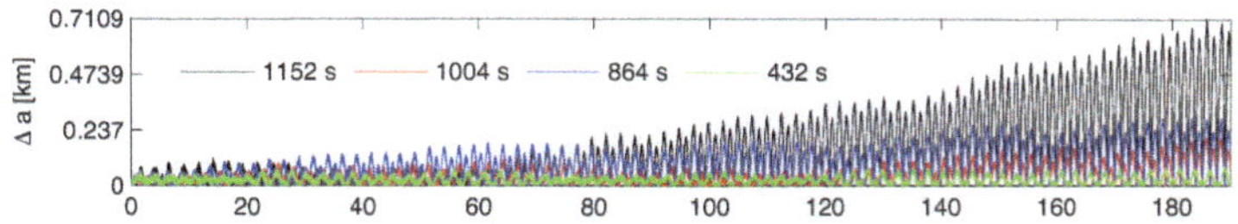

Fig. 8 Absolute differences in semi-major axis between the symplectic ($SABA_4$) and non symplectic integrators (NIMASTEP with ABM_{10}). Initial conditions are $a = 42164.140$ km, $e = 0.1$, $i = 5^\circ$, $\Omega = \omega = M = 0^\circ$. The model includes the geopotential up to degree and order 4, the solar radiation pression (with $A/m = 0.01\, m^2/kg$) and luni-solar perturbations. The initial JD is 2455194.5 days. From [9]

2.5 *The shadowing effects*

If the solar radiation pressure plays an important role in the dynamics, as for the large A/m coefficients, the crossing of the Earth shadow can not be neglected anymore and has to be taken into account. However a simple switch (on/off) in the numerical integration could be a source of numerical errors and is not compatible with a symplectic scheme.

A nice mathematical approach has been proposed [9] replacing the crossing of the umbra by smooth functions adapted to two situations: the cylindrical geometry, where the Sun is assumed to be at infinity, and the conical geometry, where the Sun is at finite distance from the Earth. Both models have been adapted to the symplectic integrator SYMPLEC. The results proved the accuracy of the conical model, with the succession of the umbra and penumbra regions (see Fig. 9), in comparison with the cylindrical one.

The shadowing effects, following the pioneer works of [8] and [1], can also be presented as perturbations acting on a reference model (two-body problem and solar radiation pressure). A semi-analytical theory (based on an analytical Hamiltonian but associated to a numerical integration of the explicit equations of motion) can be deduced, showing the existence of a very long periodic effect (several thousands of years) connected with the shadowing seasons. Even if this effect is much less important than the J_2 periodic contribution on the semi-major axis, it is well present and could appear as a fundamental period in a frequency analysis.

The introduction of the shadowing effects in the model contributes to increase the presence of chaos in the integrations, especially for high A/m coefficients. This regular passage through the umbra perturbs most of the orbits and makes the motion unpredictable after a few years, as a MEGNO analysis has clearly showed [11].

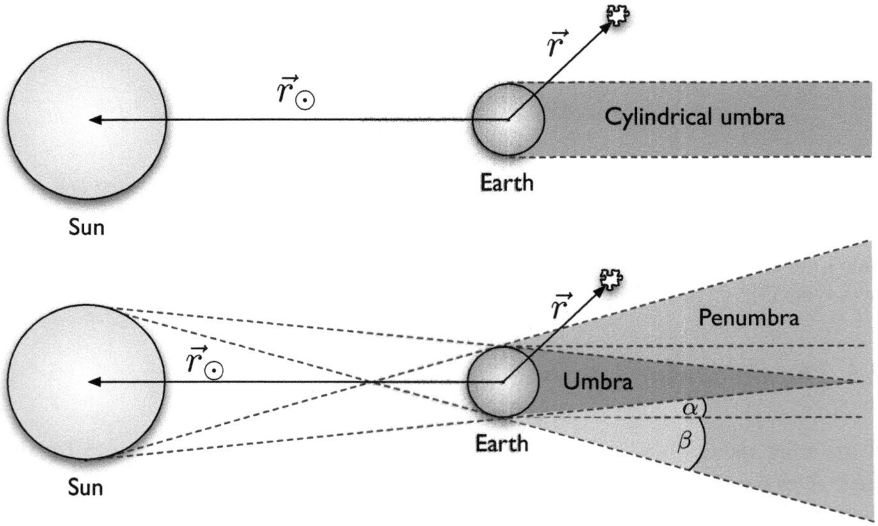

Fig. 9 Cylindrical and conical geometry to model the passage of space debris in the Earth shadow. From [9]

3 Conclusion

The mathematics, and more specifically the celestial mechanics, play an important role in the understanding of the space debris dynamics. This uncontrolled population has to be treated as a collection of bodies, and followed, modeled, anticipated on long periods of time, mixing statistical and deterministic approaches. The active debris removal (ADR) strategies are not only technological challenges, but will also require imagination and creativity, rigorous and reliable previsions. The mathematical community has an important role to play, to protect and to build the planet Earth in the future.

References

1. Aksnes, K.: Short-period and long-period perturbations of a spherical satellite due to direct solar radiation. Celestial Mechanics **13**, 89–104 (1976)
2. Beutler, G.: Methods of celestial mechanics. Physical, mathematical, and numerical problems, Volumes **1 & 2**. Springer-Verlag, Berlin Heidelberg (2005)
3. Breiter, S., Wytrzyszczak, I., Melendo, B.: Long-term predictability of orbits around the geosynchronous altitude. Adv. Space Res. **35**, 1313–1317 (2005)
4. Chao, C.C.: Analytical Investigation of GEO Debris with High Area-to-mass Ratio, AIAA paper No. AIAA-2006-6514. Presented at the 2006 AIAA/AAS Astrodynamics Specialist Conference. Keystone, Colorado (2006)
5. Cincotta, P. M., Simó, C.,: Simple tools to study global dynamics in non-axisymmetric galactic potentials, Astronomy and Astrophysics. Supplement **147**, 205–228 (2000)

6. Cincotta, P.M., Giordano, C.M., Simó, C.: Phase space structure of multi-dimensional systems by means of the mean exponential growth factor of nearby orbits. Physica D **182**, 151–178 (2003)
7. Delsate, N., Compère, A.: NIMASTEP: a software to modelize, study and analyze the dynamics of various small objects orbiting specific bodies. Astronomy & Astrophysics **540**, A120 (2012)
8. Ferraz-Mello, S.: Analytical study of the Earth's shadowing effects on satellite orbits. Celestial Mechanics **5**, 80-101 (1972)
9. Hubaux, Ch., Lemaître, A., Delsate, N., Carletti, T.: Symplectic integration of space debris motion considering several Earth's shadowing models. Adv. Space Res. **49**, 1472–1486 (2012)
10. Hubaux, Ch., Lemaître, A.: The impact of Earth's shadow on the long-term evolution of space debris. Celest. Mech. & Dyn. Astron. **116**, 79–95 (2013)
11. Hubaux, Ch., Libert, A.-S., Delsate, N., Carletti, T.: Influence of Earth's shadowing effects on space debris stability. Adv. Space Res. **51**, 25–38 (2013)
12. Laskar, J., Robutel, P.: High order symplectic integrators for perturbed Hamiltonian systems. Celest. Mech. & Dyn. Astron. **80**, 39–62 (2001)
13. Laskar, J.: Frequency analysis of a dynamical system. Celest. Mech. & Dyn. Astron. **56**, 191–196 (1993)
14. Lemaître, A., Delsate, N., Valk, S.: A web of secondary resonances for large A/m geostationary debris. Celest. Mech. & Dyn. Astron. **104**, 383–402 (2009)
15. Rossi, A.: Population models of space debris. In: Knezevic, Z., Milani, A. (eds) Dynamics of population of planetary systems. Proceedings of the IAU Colloquium No. 197, 427–438 (2005)
16. Rossi, A.: Resonant dynamics of Medium Earth Orbits: space debris issues. Celest. Mech. & Dyn. Astr. **100**, 267–286 (2008)
17. Schildknecht, T., Früh, C., Herzog, A., Hinze, J. & Vananti, A.: AIUB Efforts to Survey, Track, and Characterize Small-Size Objects at High Altitudes. Proceedings of 2010 AMOS Technical Conference, 14-17 September. Maui, Hawaii, USA (2010)
18. Tommei, G., Milani, A. & Rossi, A.: Orbit determination of space debris: admissible regions. Celest. Mech. & Dyn. Astr. **97**, 289–30 (2007)
19. Valk, S., Lemaître, A., Anselmo, L.: Analytical and semi-analytical investigations of geosynchronous space debris with high area-to-mass ratios. Adv. Space Res. **41**, 1077–1090 (2008)
20. Valk, S., Lemaître, A., Deleflie, F.: Semi-analytical theory of mean orbital motion for geosynchronous space debris, under gravitational influence. Adv. Space Res. **43**, 1070–1082 (2009)
21. Valk, S., Lemaître, A.: Semi-analytical investigations of high area-to-mass ratio geosynchronous space debris including Earth's shadowing effects, Adv. Space Res. **42**, 1429–1443 (2008)
22. Valk, S. Delsate, N., Lemaître, A., T.: Global dynamics of high area-to-mass ratios GEO space debris by means of the MEGNO indicator Adv. Space Res. **43**, 1509–1526 (2009)
23. Yoshida, H.: Construction of higher order symplectic integrators. Phys. Lett. A **150**, 262–268 (1990)

Mathematical Models for Socio-economic Problems

Maria Letizia Bertotti and Giovanni Modanese

Abstract We discuss a framework for the microscopic modelling of taxation and redistribution processes in a closed trading market society. For a prototype model and some variants of it, we examine the emergence of income distribution curves which exhibit "fat" power-law tails as the real world ones. We also incorporate tax evasion into the models and we investigate, in particular, its effect on the income profiles. Our findings are in agreement with the expectation that a fair fiscal policy and individual correctness are effective tools towards the overcoming of social inequalities.

1 Introduction and some historical perspective

Issues like wealth inequality and the income distribution in a population or the effects of tax evasion certainly fall within the range of subjects of concern for us humans, inhabitants of the Planet Earth. Such issues have been the object of interest virtually since ever. But, especially today, also thanks to the increased power of computer simulations can mathematics help understand and explain the dynamics of the emergence of collective patterns in complex socio-economic systems. In fact, by providing suitable frames for modelling, analytical methods and computational tools, mathematics can contribute to the exploration of possible scenarios, sometimes

---

M. Letizia Bertotti (✉)
Faculty of Science and Technology, Free University of Bozen-Bolzano, Piazza Università 5, 39100 Bolzano, Italy
e-mail: MariaLetizia.Bertotti@unibz.it

G. Modanese
Faculty of Science and Technology, Free University of Bozen-Bolzano, Piazza Università 5, 39100 Bolzano, Italy
e-mail: Giovanni.Modanese@unibz.it

A. Celletti, U. Locatelli, T. Ruggeri and E. Strickland (eds.): *Mathematical Models and Methods for Planet Earth.* Springer INdAM Series 6, DOI 10.1007/978-3-319-02657-2_10,

even suggesting to policy makers conceivable actions or welfare measures. Based on these considerations, the above mentioned topics seem to us to deserve a place under the theme "A planet organised by humans" in the MPE2013 program, see [1].

What we want to emphasise here is a particular perspective through which socio-economic questions as those mentioned above, but more generally also mechanisms underlying financial markets and similar matters, can be analysed. This perspective began to take shape, during the last decades. Unlike traditional views, it puts the interactions among heterogeneous individuals at the very heart of the question. In other words, it recognizes that these interactions play a crucial role towards the self-organisation of the systems and the formation of the observable aggregate features. This is quite different from the traditional viewpoint of mainstream economics, whose cornerstones are the assumption of a representative agent and the rational choice theory. According to the novel perspective, the behaviour at the collective level is not and cannot be that of an average individual, but rather results from the different individual peculiarities and their interplay.

A survey on this area and in this direction, especially one describing and comparing approaches of researchers from different disciplines, would take a number of pages, also in view of the multiplicity of the specific issues addressed. To give at least some references, we recall e.g. that arguments supporting the need of a behaviour- and interaction-based approach to economics and finance can be found in the recent book by A. Kirman [16]. Pioneers in this direction were, among others, T. Schelling and B.W. Arthur. Two descriptive and insightful readings are e.g. [2] and [19].

The typical methodology employed in connection with the mentioned paradigm primarily includes computational agent-based models [13]. Network theory also provides additional important ingredients since the structure of the relationships or links between individuals clearly has a significant influence on the dynamics of socio-economic processes. With regard to the tax evasion problem, agent-based models and simulations have been developed and carried out e.g. in [8, 14, 22]. In these papers, behaviourally different agent types, such as honest, imitative, tax evaders and so on, are considered. The focus is on the effect of individual interactions and enforcement mechanisms, including auditing and punishment procedures, on tax compliance and on changes in individual behaviour patterns. A similar goal drives the experimental approach described in [18].

On the other side, bottom-up approaches to the formation of the income distribution have been investigated e.g. in [9–11, 20, 21], mainly by authors belonging to the physics community. The reason for the presence of physicists among researchers dealing with social and economic topics is that these topics refer to systems (populations) composed by a large number of interacting elements (individuals). It is therefore natural to investigate them by means of tools which were originally developed in statistical mechanics and in gas kinetic theory. Already L. Boltzmann had suggested this as a possible route to be explored some day. It was more than a century ago, in a conference held at the 1904 St. Louis Congress:[1]

> This opens a broad perspective if we do not only think of mechanical objects. Let's consider to apply this method to the statistics of living beings, society, sociology and so forth.

The denomination "econophysics" was coined by H.E. Stanley in the mid – 1990s to designate the research field which explores the dynamical behaviour of economic and financial markets employing statistical physics methods, [17].

The approach which we suggest has a mathematical character. It relies on a general framework for the study of monetary exchanges in a closed market society in the presence of taxation and redistribution. This framework was introduced in [4] and was then exploited also in [6] and [7] towards the construction and investigation of some specific models. Its roots lie in a discrete version, developed in [5] and subsequent papers, of a generalized kinetic theory proposed in [3] and inspired by [15]. In fact, the structure introduced in [4] is a further development of the one in [5], modified so as to take into account particular features of the novel situation. It is expressed by a system of nonlinear ordinary differential equations and it describes the time evolution of the income distribution, based on the knowledge of individual interactions.[2] It comprises as many equations as are the classes characterized by average income, in which one divides a population. Each equation gives the variation in time of the fraction of individuals belonging to a certain class. The framework contains various parameters, to be chosen on the basis of phenomenological observations. It enjoys a great flexibility. It allows, for instance, a very natural definition of different taxation rates for different income classes and a division of the population in classes whose incomes increase nonlinearly, what enables a better representation of the "super-rich" classes.

Below, we will briefly discuss this framework. Then, by implementing particular choices of some parameters, we will restrict our attention to some specific models. We will then go over the various features of these models and, in particular, we will report on the emergence of aggregate patterns in the income distribution curves, which exhibit qualitative properties like those observed in the real world. Then, as an extension of our previous work, we will incorporate into the models also the occurrence of tax evasion and we will draw some conclusion on its effect.

2 A general framework

A sketch of the mechanism which motivates the framework follows. For the sake of conciseness, we avoid introducing here the more general case.

---

[1] The text of the conference can be found under the title "The Relations of Applied Mathematics" e.g. in the volume *Physics for a New Century. Papers presented at the* 1904 *St. Louis Congress*, edited by Katherine Russell Sopka, Tomash Publishers/Amer. Inst. of Physics (1986), pp. 267–279.

[2] As will be clear in the sequel, we consider here also interactions with a nonlinear adaptive nature. Moreover, a single interaction does not lead here to a change of class of individuals as was the case in [5].

Divide a population of individuals into a finite number n of classes, each one characterized by its average income. Let $r_1 < r_2 < \ldots < r_n$, denote the average incomes of the n classes, and let $x_i(t)$, where $x_i : \mathbf{R} \to [0, +\infty)$ for $i \in \Gamma_n = \{1, 2, \ldots, n\}$, denote the fraction at time t of individuals belonging to the i-th class. In the following, the indices i, j, h, k, etc. always belong to Γ_n, if no differently stated. Assume that pairwise interactions of economic nature, subjected to taxation, take place. And call S the fixed amount of money that people may exchange during their interactions.

Any time an individual of the h-th class has to pay a quantity S to an individual of the k-th class, this one in turn has to pay some tax corresponding to a percentage of what he is receiving. This tax is quantified as $S\tau$, with the tax rate $\tau = \tau_k \leq 1$ depending in general on the class of the earning individual. Since the quantity $S\tau$ goes to the government, which is supposed to use the money collected through taxation to provide welfare services for the population, we interpret the welfare provision as an income redistribution. Ignoring the passages to and from the government, we may imagining that equivalently: in correspondence to any interaction between an h-individual and a k-individual, where the one who has to pay S to the other one is the h-individual, this pays to the k-individual a quantity $S(1-\tau)$ and he pays as well a quantity $S\tau$, which is divided among all j-individuals for $j \neq n$.[3] Accordingly, the effect of taxation and redistribution is equivalent to the effect of a quantity of interactions between the h-individual and each one of the j-individuals for $j \neq n$, which are induced by the effective h-k interaction. To fix notations, we may distinguish between *direct* interactions (h-k) and *indirect* interactions (h-j for $j \neq n$). Any direct or indirect economical interaction yields as a consequence a possible slight increase or slight decrease of the income of individuals.

All this can be expressed through the system of the n differential equations

$$\frac{dx_i}{dt} = \sum_{h=1}^{n} \sum_{k=1}^{n} \left(C_{hk}^i + T_{[hk]}^i(x) \right) x_h x_k - x_i \sum_{k=1}^{n} x_k \,, \qquad i \in \Gamma_n \,, \tag{1}$$

where the *direct transition probability densities* $C_{hk}^i \in [0, +\infty)$, expressing the probability density that an individual of the h-th class will belong to the i-th class after a direct interaction with an individual of the k-th class, satisfy $\sum_{i=1}^n C_{hk}^i = 1$ for any fixed h and k and the *indirect transition variation densities* $T_{[hk]}^i : \mathbf{R}^n \to \mathbf{R}$ account for the indirect interactions and express the variation density in the i-th class due to an interaction between an individual of the h-th class with an individual of the k-th class. The $T_{[hk]}^i(x)$ with $x = (x_1, \ldots, x_n) \in \mathbf{R}^n$ are continuous functions and satisfy $\sum_{i=1}^n T_{[hk]}^i(x) = 0$ for any fixed h, k and $x \in \mathbf{R}^n$.

[3] The reason why individuals of the n-th class constitute an exception is a technical one: if an individual of the n-th class would receive some money, the possibility would arise for him to advance to a higher class, which is impossible.

3 A family of models. Analytical results

To design within the general framework (1) a specific model, the expressions of the C^i_{hk}'s and the $T^i_{[hk]}(x)$'s need to be further characterized. We follow here the path of [4,6,7]. Referring to these papers for more details, we take $C^i_{hk} = a^i_{hk} + b^i_{hk}$, where the only nonzero elements a^i_{hk} are $a^i_{ij} = 1$ for $i, j \in \Gamma_n$ and the only possibly nonzero elements b^i_{hk} are of the form

$$b^i_{i+1,k} = p_{i+1,k}\, S\, \frac{1-\tau_k}{r_{i+1}-r_i},$$

$$b^i_{i,k} = -\,p_{k,i}\, S\, \frac{1-\tau_i}{r_{i+1}-r_i} - p_{i,k}\, S\, \frac{1-\tau_k}{r_i - r_{i-1}},$$

$$b^i_{i-1,k} = p_{k,i-1}\, S\, \frac{1-\tau_{i-1}}{r_i - r_{i-1}}. \tag{2}$$

The elements $p_{h,k}$ appearing in (2) express the probability that in an encounter between an h-individual and a k-individual, the one who pays is the h-individual. If the possibility is admitted that the two individuals do not interact, the $p_{h,k}$ are required to satisfy $0 \le p_{h,k} \le 1$ and, of course, $p_{h,k} + p_{k,h} \le 1$. Apart from that, they can be chosen with a certain arbitrariness. The expressions in (2), as well as similar ones in the following, are defined only for meaningful values of the indices.

We take $T^i_{[hk]}(x) = U^i_{[hk]}(x) + V^i_{[hk]}(x)$, where the variation density corresponding to the advancement from a class to the subsequent one, due to the benefit of taxation is

$$U^i_{[hk]}(x) = \frac{p_{h,k}\, S\, \tau_k}{\sum_{j=1}^n x_j} \left(\frac{x_{i-1}}{r_i - r_{i-1}} - \frac{x_i}{r_{i+1} - r_i} \right) \tag{3}$$

and the variation density corresponding to the retrocession from a class to the preceding one, due to the payment of some tax is

$$V^i_{[hk]}(x) = p_{h,k}\, S\, \tau_k \left(\frac{\delta_{h,i+1}}{r_h - r_i} - \frac{\delta_{h,i}}{r_h - r_{i-1}} \right) \frac{\sum_{j=1}^{n-1} x_j}{\sum_{j=1}^n x_j}. \tag{4}$$

As usual, $\delta_{h,k}$ denotes the *Kronecker delta*.

A theorem proved in [4] ensures that, in correspondence to any initial condition $x_0 = (x_{01}, \dots, x_{0n})$, for which $x_{0i} \ge 0$ for all $i \in \Gamma_n$ and $\sum_{i=1}^n x_{0i} = 1$, a unique solution $x(t) = (x_1(t), \dots, x_n(t))$ of (1) exists, which is defined for all $t \in [0, +\infty)$, satisfies $x(0) = x_0$ and also

$$x_i(t) \ge 0 \text{ for } i \in \Gamma_n \text{ and } \sum_{i=1}^n x_i(t) = 1 \text{ for all } t \ge 0. \tag{5}$$

We emphasize that in fact the only meaningful initial data x_0 and solutions $x(t)$ are those with non negative components, whereas the constraint $\sum_{i=1}^n x_{i0} = 1$ simply expresses a normalization. Ultimately, this well-posedness result guarantees that the solutions of interest are distribution functions and allows to further simplify the expressions of the $U^i_{[hk]}(x)$ and $V^i_{[hk]}(x)$ in (3) and (4). They become linear in the variables x_j and, accordingly, the right hand sides of the Equations (1) are polynomials containing cubic terms as the highest degree ones. As proven in [4], the scalar function $\mu(x) = \sum_{i=1}^n r_i x_i$, expressing the global income and, due to the population normalization, also the mean income, is a first integral for the system (1).

Due to the fact that the value of n and the parameters $r_k, \tau_k, p_{h,k}$ are still to be fixed, the Equations (1) actually describe a family of models rather than a single model.

4 A specific model and related income distribution profiles

In order to investigate the long-time behaviour of the solutions of the equations (1), we have to fix the parameter values. While noticing that in [4,6,7] also other parameter choices are discussed, we take here $n = 25$, and we choose $r_j = 10j$ and

$$\tau_j = \tau_{min} + \frac{j-1}{n-1}(\tau_{max} - \tau_{min}) \tag{6}$$

for $j \in \Gamma_n$, where $\tau_{min} = 30\%$ and $\tau_{max} = 60\%$. Anyway, we will also mention different choices of r_j, τ_{min} and τ_{max}. As for the $p_{h,k}$'s, we choose

$$p_{h,k} = (1/4)\min\{r_h, r_k\}/r_n$$

with the exception of the terms

$$\begin{aligned}
p_{1,k} &= 0 && \text{for } k \in \Gamma_n\,, \\
p_{h,n} &= 0 && \text{for } h \in \Gamma_n\,, \\
p_{h,1} &= (1/2)\, r_1/r_n && \text{for } h \in \{2,...,n\}\,, \\
p_{j,j} &= (1/2)\, r_j/r_n && \text{for } k \in \{2,...,n-1\}\,, \\
p_{n,k} &= (1/2)\, r_k/r_n && \text{for } k \in \{1,...,n-1\}\,.
\end{aligned}$$

This choice expresses a certain heterogeneity in the saving propensity of individuals of different classes. We also take $S = 1$.

The resulting equations are well beyond analytical solutions. But, several simulations give evidence of the facts outlined in the following subparagraphs.

Uniqueness of the asymptotic stationary distribution for any fixed value of μ

For any fixed value $\mu \in [r_1, r_n]$ of the global income, an equilibrium of (1) – namely a stationary distribution - exists, which coincides with the asymptotic trend of all solutions of (1), whose initial conditions $x_0 = (x_{01}, \ldots, x_{0n})$ satisfy $x_{0i} \geq 0$ for all $i \in \Gamma_n$, $\sum_{i=1}^n x_{0i} = 1$, and $\sum_{i=1}^n r_i x_{0i} = \mu$. In other words, a one-parameter family of asymptotic stationary distributions exists. We stress here that what remains constant in these stationary distributions is the ratio of individuals in each class and not the single individuals. Those continue to be susceptible to move from a class to another.

Dependence of the asymptotic stationary distribution on the difference between the maximum and the minimum tax rate

It is worthwhile also comparing the output of simulations relative to the model at hand and others belonging to the same family. We refer here to models in which n and the parameters $r_k, p_{h,k}$ and μ, as well as the formula (6), are as above, but different values of τ_{min} and τ_{max} are taken. What one observes in these cases is that the profile of the asymptotic stationary distribution depends on the difference between the rates respectively applied to the highest and to the lowest income classes. Specifically, to an increase of the difference $\tau_{max} - \tau_{min}$ between the maximum and the minimum tax rates, a growth of the middle classes at the asymptotic equilibrium corresponds, to the detriment of the poorest and the richest classes (see also [4]). Also the dependence on the conserved quantity μ might be investigated. Results for different models can be found in [6].

Emergence of power-law distribution tails

The most intriguing property we want to highlight here is the emergence, for suitable values of μ, of a power-law decreasing behaviour in the higher income section of the asymptotic stationary distributions tails. We emphasize that towards this goal we restricted attention to initial conditions, i.e. initial distributions of the population, having the majority of individuals concentrated in lower income classes.[4] Of course, these are the typical situations occurring in real world societies.

To give an overall picture, the asymptotic stationary distributions exhibit, for suitable values of μ, the following patterns: the density of the low income classes is smaller than for the low-medium classes, the maximal density is achieved by the low-medium classes and the density progressively decreases for the higher income classes, according to a power-law. The reason of interest of this output is that it is known since Pareto discovered it more than a century ago, and it's also confirmed by recent empirical data, see e.g. [10], that in several different countries and in different times, the high-end of the income distributions follow a power-law.

The power-law behaviour of a function $f(x) = cx^{-\alpha}$ is equivalently expressed by the equation $\ln f(x) = \ln c - \alpha \ln x$. Therefore, if the distribution tail exhibits such a behaviour, in the log-log plot in the variables x and f a straight line must be seen. The number $\beta = \alpha - 1$ is known as Pareto index and was assessed by Pareto itself to be approximately equal to 1.5. More generally, β was observed to take values between 1 and 2.

We observe here that a power-law tail would be more clearly distinguished in correspondence to an income range of at least two magnitude orders. In this connection, we also investigated, besides linear class incomes, incomes which increase exponentially as $r_i = c\,(1,67)^i$ for some positive constant c. The basis 1.67 is chosen in such a way as to obtain an income increase by a factor 100 when the class index i increases by 9 (the tail of the distribution can be taken as the part relative to

[4] Since the number of individuals is constant in time and only a finite number of income classes is scheduled, if a tail is expected in the asymptotic distribution, the global income cannot be too high.

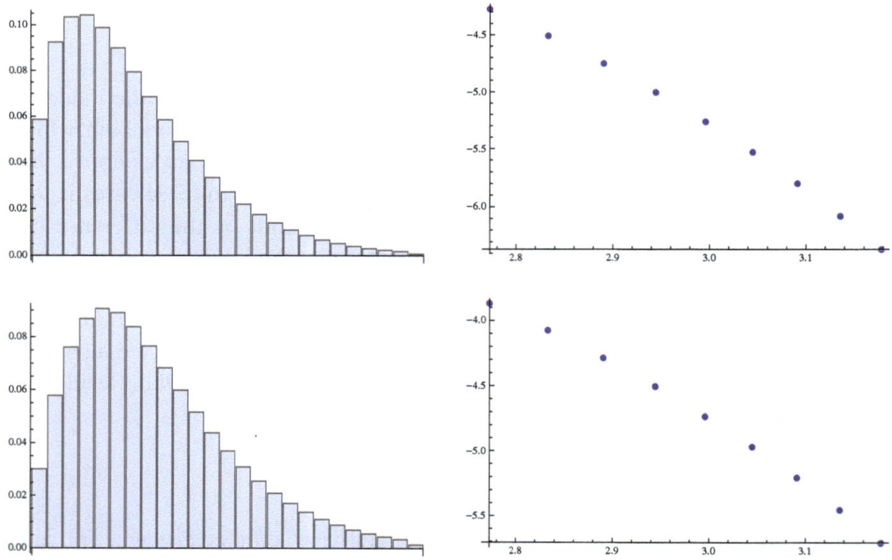

Fig. 1 In each of the two rows in the figure, the panels on the left show the histograms corresponding to an asymptotic stationary distribution, and those on the right display the log-log plots of the right part of the distributions. We warn that the histograms are scaled differently from panel to panel

the classes from $i = 16$ to $i = 25$). A rescaling of the discrete distribution function x_i is then necessary. In this case, the rescaling factor turns out to be proportional to the reciprocal income r_i^{-1}, so that that a power law tail maintains, after rescaling, the power-law form, with a Pareto index equal to the index of the x_i augmented by 1.

As a sample among many similar ones, Figure 1 illustrates (a discrete version of) the power-law behaviour of the asymptotic distributions of the model with $\tau_{min} = 45\%$ and $\tau_{max} = 60\%$ for two different values of the global income μ.

We presume that the main ingredient for the power-law tail emergence in our model is provided by the heterogeneity of the saving propensity of individuals. For instance, if a transaction between a poor and a rich guy takes place, in general (except for individuals of the poorest and the richest class) the quantity of money exchanged is the same, independently of who is paying. Indeed, the effect of the mechanism described above is the same as if, when trading, both individuals would pay a small quantity proportional to the income of the poorest of the two. Accordingly, the saving propensity of rich people is higher. We may conclude that the model entails a kind of "protection" mechanism, which allows rich people to accumulate money and to form a "fat" tail. In addition, possibly, also the lack of time-reversal symmetry plays a role.

5 A model with tax evasion

In this section we incorporate in the model a partial tax evasion occurrence and analyze the difference, in correspondence to the same initial conditions, between the asymptotic income distribution curves obtained in the presence of this illegal practice and the income distribution curves obtained in the presence of tax compliance.

There are in the real world different forms of tax evasion. Here, we restrict our attention to the case in which this illegal practice provides an advantage not only to the individual who on the occasion of a trade is earning money, but also to the one who is paying.[5] This may happen for example in connection with value added tax (VAT). The payment of such a tax is largely based upon receipts and invoices. Accordingly, typical features of the related tax evasion can be summarized as follows.

Let the individual who receives the money, for instance a trader, a professional, an entrepreneur, and so on, be a k-individual. He may collude with the individual who is paying him, a h-individual, offering a discount in exchange for the agreement that no issuance of a receipt or invoice will be required. Then, the advantage for the h-individual consists in the discount, while the advantage for the k-individual is that, since there is no trace of the transaction, the gain can be concealed from the tax return and this means tax evasion.

In order to incorporate the occurrence, at least to some extent, of such a behaviour, we first recall that according to Section 2, in the case of tax compliance, when an h-individual is supposed to pay an amount of money S to a k-individual, what equivalently happens is that:

- The h-individual pays a quantity $S(1-\tau_k)$ to the k-individual and he pays to the government a quantity $S\,\tau_k$.

Now, let $\theta_k < \tau_k$. To fix ideas, we took in particular

$$\theta_k = \frac{4}{5}\,\tau_k$$

when running the simulations, on which we are going to report here. Of course, also this parameter could be variously tuned. The effect of a partial tax evasion can be produced provided e.g. we postulate that, when an h-individual is supposed to pay an amount of money S to a k-individual, as a matter of fact:

- The h-individual pays a quantity $S(1-(\tau_k+\theta_k)/2)$ to the k-individual and he pays to the government a quantity $S\,\theta_k$.

In this way, the h-individual pays less than he should, the k-individual gains in the end more than he would have done in the tax compliant situation and the government collects less than it should.

As already announced, our aim here is to compare the asymptotic income distribution curves in the tax compliance case with those in the presence of tax evasion. Towards this aim, we performed a relevant number of simulations, relative to pairs

[5] We are taking into account also other evasion forms in work in progress.

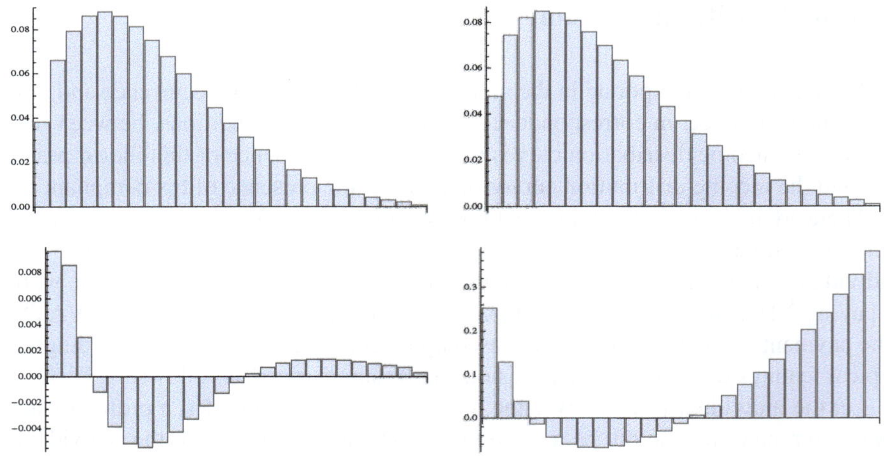

Fig. 2 In the panel on the left in the first row an asymptotic distribution in the presence of tax compliance is showed; the panel on the right in the first row displays the asymptotic distribution for the same initial condition in the presence of tax evasion; the histograms in the panel on the left in the second row express the difference of the fraction of individuals in each class in the cases with tax evasion and with tax compliance; finally, the histograms in the panel on the right in the second row represent the percentage of variation in each class of the fraction of individuals when passing to a tax compliance case to one with tax evasion. The histograms are scaled differently on different pictures

of cases, the first one comprising tax compliance and the other one with tax evasion, having the same initial conditions.

The simulations systematically show that the effect of the tax evasion is that there is an increment in the number of individuals belonging to the poorest classes and the richest ones at the detriment of the middle classes. It is also of interest to look at the percentage in each class of the variation of the number of individuals, when passing from the straight case to the dishonest case. Indeed, the increasing effect in the high income classes may be observed to be larger for greater income; vice versa, in the low income classes, it is larger for lower income. Correspondingly: those who benefit from tax evasion are individuals of the richest classes; and the richer they are, the most they benefit. In contrast, the situation of individuals having low income is getting worse: many of them even pass to the lowest income class. The four panels in the Figure 2 illustrate the typical output.

We also point out that the Gini index turns out to be larger when tax evasion is present. This index provides a well known measure of inequality (in the present context) of income. It can range from the value 0 for complete equality to the value 1 for the maximal inequality. It is obtained based on the Lorenz curve, which plots the cumulative percentage of the total income of a population (on the y axis) earned by the bottom percentage of individuals (on the x axis). In comparison with the Lorenz curve, the line at 45 degrees represents a perfect equality of incomes. The Gini index is defined as the ratio of the area between the Lorenz curve and the line of perfect equality and the total area under the line of perfect equality. In the examples in our simulations we estimated the Gini index by calculating the area under the Lorenz curve as a sum of areas of trapezia.

6 Conclusions

We discussed here a framework for the microscopic modelling of a closed trading market society and for the description of the processes of taxation and redistribution. We focused our attention on a prototype-model (and some variations of it) and for that model we examined the emergence, starting from different sets of initial conditions, of the income distribution curve. In particular, we investigated the effect on this curve of tax evasion. The specific model we built is certainly rudimentary; however, it apparently captures some relevant ingredients of the problem. In a nutshell, it produces income profiles with fat tails as in the real world, and it is in agreement with the belief that a fair fiscal policy and individual correctness are effective tools for the overcoming of social inequalities. Beyond this particular model, we emphasize that our primary goal was to build a general frame suitable to establish the correspondence between microscopic interactions and macroscopic patterns and to lead to qualitatively reasonable results.

Before concluding, we just like to add a couple of remarks.

Plainly, the strategy of admitting a finite number of income classes entails the introduction of some unnatural constraint in the problem. But, we think that in an application-oriented perspective it is not so out of place: typically, in reports and studies concerning real world populations, income distributions are not infinitely extended, and often they are described by a finite number of income bins. Besides, statements concerning the power law decreasing behaviour are not to be intended in a strictly analytical way.

When running simulations for models defined by different choices of the parameters $p_{h,k}$, one immediately realizes that these choices may have a significant effect on the dynamics of the relative systems. And as a matter of fact, the choice of the $p_{h,k}$ is characterised by a high degree of arbitrariness. This point deserves a deeper exploration: a calibration of these and other parameters more related to empirical data is in order.

Somehow related is the problem of defining a network of connections. Of course, frequency encounter rates between individuals of any given class with individuals of another given class can be easily introduced. Instead, the models designed within the framework at hand do not seem to be suitable to be equipped with networks having single individuals as nodes: all individuals of a same class are considered to be equivalent as far as their interaction behaviour with other individuals is concerned.

There are henceforth a number of further aspects and directions towards which the investigation can be addressed. Among several ones motivated by an interest of applicative nature, also the challenge of constructing analytical proofs of certain computationally evident facts is attractive and stimulating.

References

1. Mathematics of Planet Earth (2013). http://www.mpe2013.org
2. Arthur, B., Durlauf, S., Lane, D.A.: Process and Emergence in the Economy. In: Arthur, B., Durlauf, S., Lane, D.A. (eds.) The Economy as an Evolving Complex System II, pp. 2–14. Addison Wesley, Reading (1997)
3. Arlotti, L., Bellomo, N., De Angelis, E.: Generalized kinetic (Boltzmann) models: mathematical structures and applications. Math. Mod. Meth. Appl. Sci. **12**, 567–592 (2002)
4. Bertotti, M.L.: Modelling taxation and redistribution: a discrete active particle kinetic approach. Appl. Math. Comput. **217**, 752–762 (2010)
5. Bertotti, M.L., Delitala M.: From discrete kinetic and stochastic game theory to modelling complex systems in applied sciences. Math. Mod. Meth. Appl. Sci. **14**, 1061–1084 (2004)
6. Bertotti, M.L., Modanese G.: From microscopic taxation and redistribution models to macroscopic income distributions. Physica A **390**, 3782–3793 (2011)
7. Bertotti, M.L., Modanese G.: Exploiting the flexibility of a family of models for taxation and redistribution. Eur. Phys. J. B **85**, 261 (2012)
8. Bloomquist K.M.: A Comparison of Agent-Based Models of Income Tax Evasion. Social Science Computer Review **24**, 411–425 (2006)
9. Chakraborti A., Chakrabarti B.K.: Statistical mechanics of money: how saving propensity affects his distribution. Eur. Phys. J. B **17**, 167–170 (2000)
10. Chatterjee A., Yarlagadda S., Chakrabarti B.K. (Eds.): Econophysics of Wealth Distributions. Springer-Verlag Italia, Milan (2005)
11. Dragulescu A., Yakovenko V.M.: Statistical mechanics of money. Eur. Phys. J. B **17**, 723–729 (2000)
12. Duering, B. Matthes, D. Toscani G.: A Boltzmann-type approach to the formation of wealth distribution curves. Riv. Mat. Univ. Parma, Series **8.1**, 199–261 (2009)
13. Gallegati M.: Reconstructing economics: Agent Based Models and Complexity. Conference Paper, Berlin, Paradigm Lost: Rethinking economics and Politics (2012). Available via http://ineteconomics.org/conference/berlin/reconstructing-economics-agent-based-models-and-complexity
14. Hokamp S. Pickhardt M.: Income tax evasion in a society of heterogeneous agents: evidence from an agent-based model. International Economic Journal **24**, 541–553 (2010)
15. Jaeger E., Segel L.A.: On the distribution of dominance in populations of social organisms. SIAM J. Appl. Math. **52**, 1442–1468 (1992)
16. Kirman A.: Complex Economics: Individual and Collective Rationality. Routledge, London (2010)
17. Mantegna R.N., Stanley H.E.: An Introduction to Econophysics. Cambridge University Press, Cambridge (2000)
18. Mittone L.: Dynamic behaviour in tax evasion: an experimental approach. The Journal of Socio-Economics **35**, 813–835 (2006)
19. Schelling T.C.: Micromotives and Macrobehavior. Revised edition (First edition in 1978). W.W. Norton e Company, New York (2006)
20. Sinha S., Chakrabarti B.K.: Towards a physics of economics. Physics News (Bullettin of the Indian Physical Association) **39**, 33–46 (2009)
21. Yakovenko V.M.: Econophysics: Statistical mechanics approach to. In: Meyers R.A. (ed.), Encyclopedia of Complexity and System Science, pp. 2800–2826. Springer, New York (2009)
22. Zaklan G., Westerhoff F., Stauffer D.: Analysing tax evasion dynamics via the Ising model. J. Econ. Interact. Coord. **4**, 1–14 (2009)

Climate as a Complex Dynamical System

Antonello Provenzale

Abstract Climate is a complex dynamical system, whose understanding is a fascinating scientific challenge which has crucial implications on our society. Like any science, the study of climate is based on data, measurements and experiments, and requires the development of analysis and modeling tools to build a coherent view of climate and its variability. In such a framework, the role of Mathematics and of mathematical developments is essential. In the present contribution, I discuss the hierarchy of climate models, mentioning both the progress of the last thirty years, the role of mathematical approaches, and the many open questions which still need to be clarified. The need for a coherent theory of climate is advocated, as a world-wide effort to understand this fascinating manifestation of Planet Earth.

1 Planetary climates

The climate of our planet varies on all time and space scales, and the Earth has undergone periods which were much warmer than today (by 10 degrees Kelvin and more), as well as times when it was definitely colder – and maybe covered with extensive ice sheets as during the so-called Snowball Earth episodes. On the other hand, the climate of our planet is also remarkably stable, and – till now – it did not push the Earth into a final frozen state or into Venus-like conditions with surface temperatures of several hundred centigrades. Stability and variability (in a limited interval): these are, probably, the two most intriguing characteristics of the Earth's climate. Stability, in particular, is the characteristics that allows for the presence of complex life, and – possibly – is at least partially a product of the presence of life [19].

The climate of our planet is a complex and complicated systems: it is complicated because it is composed of many different sub-systems (atmosphere, ocean,

A. Provenzale (✉)
Institute of Atmospheric Sciences and Climate, National Research Council, Corso Fiume 4, I-10133 Torino, Italy
e-mail: a.provenzale@isac.cnr.it

A. Celletti, U. Locatelli, T. Ruggeri and E. Strickland (eds.): *Mathematical Models and Methods for Planet Earth.* Springer INdAM Series 6, DOI 10.1007/978-3-319-02657-2_11,

cryosphere, soil, biosphere, and more recently anthroposphere), and it is complex because these different components interact with each other in a nonlinear way, on several space and time scales, and create many positive and negative feedback loops which can amplify or moderate perturbations [24,25]. In addition, Earth's climate is subject to external forces, which also vary in space and time and induce changes to which climate often responds in a nonlinear way.

The driving force of climate is solar radiation, without which there would be no climate at all. Solar energy warms the planet surface, which re-emits infrared radiation until an equilibrium is reached. Using the first law of Thermodynamics and the Stefan-Boltzmann law, we can equate the incoming and outgoing energy fluxes and obtain the surface temperature of the planet. Since the Earth is spherical and rotates on itself, by averaging over one rotation period (that is, one day) we obtain

$$S(1-A)/4 = \sigma T^4, \tag{1}$$

where T is the surface temperature of the planet in Kelvin, $S = 1367$ W m^{-2} (on average) is the solar constant, $\sigma = 5.67 \cdot 10^{-8}$ W m^{-2} K^{-4} is the Stefan-Boltzmann constant, and the factor 1/4 is the ratio between the cross section of the Earth (πR^2 where R is the Earth's radius) and the surface over which the heat is distributed ($4\pi R^2$). The parameter A is the albedo of the planet, that is, the fraction of solar energy which is reflected without entering the thermal machinery of planetary climate (for the Earth as a whole, $A \approx 0.3$). The simple equation written above provides, for our planet, a value of about -18 degree Celsius, very different from the true average temperature of the planetary surface, which is about 15 degree Celsius.

The reason for this discrepancy is that the Earth's atmosphere is capable of partially absorbing the outgoing infrared radiation (while it is largely transparent to the incoming visible radiation), generating a greenhouse effect. The absorbed infrared radiation is re-emitted and about half of the re-emitted radiation reaches the planet surface, making it warmer than predicted by equation (1), while the upper portion of the atmosphere becomes cooler. Planets without an atmosphere would not have any greenhouse effect, and equation (1) would produce the correct surface temperature.

The fact that the Earth is spherical then generates a non-uniform distribution of solar heating on the planet surface: in our case, with the axis of rotation almost perpendicular to the planetary orbit, the polar regions receive less energy than the equatorial regions. On a planet without fluid envelopes (that is, ocean or atmosphere), this would generate strong temperature differences between the Equator and the Poles, and there would be a local balance between incoming and outgoing radiation. On Earth, however, there are fluids which can move and transport heat from the equatorial regions to the Poles, leading to a weaker temperature difference across latitudes. This differential heating is at the very heart of the climate system: almost all fluid motions in the atmosphere and the ocean (except tides and geothermally-induced motions) are driven, one way or another, by the differential solar heating of the planet. Clearly, the existence of a fluid envelope is a crucial ingredient to have climate dynamics at all.

The amount of solar radiation reaching the Earth's surface varies with time. The fact that the Earth rotates rapidly with respect to its orbital period allows for exposing the whole planetary surface to solar radiation over one day (climate would be very different for a tidally-locked planet). Thus, for most climatic purposes we can average the incoming solar radiation over one day and ignore the insolation differences between night and day. Then, the fact that the rotation axis is not exactly perpendicular to the orbital plane introduces a seasonal modulation in the amount of solar radiation received at different latitudes, which has important consequences on climate. On much longer time scales, the variations in the Earth's orbital parameters induce slight variations in the amount and distribution of solar radiation reaching the planet, with potential implications on long-term climatic variations such as the alternance between glacial and interglacial periods [20].

The amount of energy emitted by the Sun could also vary in time (albeit recent measurements indicate a very weak variation during the 11-yr solar cycle), and so does also the chemical composition of the atmosphere. Large volcanic eruptions can emit huge quantities of aerosols in the atmosphere, which can reflect solar radiation and have a cooling effect on climate. In the last 150 years, human activities have modified the composition of the atmosphere, with the emission of large amounts of carbon dioxide and methane (gases with significant greenhouse effect), as well as with the emission of aerosols (which, depending on composition and color, can have a cooling or warming effect) and changes in the surface properties and land use. All these anthropic effects represent a new forcing factor for the climate of our planet.

External forcing, however, is not the whole story. Climate varies owing to its internal dynamics and to the interactions between its different components. Most climatic processes are nonlinear and can generate multiple equilibria, oscillations, and chaotic behavior. The atmosphere, by itself, can generate a complex dynamics on relatively short time scales (from days to years), including the vagaries of weather and the rich structure of planetary circulations. Even planets with a thin atmosphere and no ocean (such as Mars) have violent and extremely interesting atmospheric dynamics.

On longer time scales, the role of the ocean becomes essential, and the interactions between the atmosphere and the ocean generate new and complicated phenomena. The El Nino – Southern Oscillation (ENSO) is one example, as well as the interplay of the atmospheric circulation with the wind-driven ocean circulation (and the deeper density-driven, or thermohaline, ocean circulation). Changes in the cryosphere (ice sheets, glaciers, snow cover, permafrost), in the vegetation and in land cover are all primary players in the great game of climate variability.

As a conceptual example of climatic feedbacks, we can mention the Charney mechanism [5], which produces multiple equilibria in semi-arid environments, as a consequence of the interplay between precipitation and vegetation. Suppose that in a dry environment, where vegetation is water-limited, a perturbation in land cover reduces the amount of vegetation. As a result, more soil (sand) will be exposed, and the albedo of the surface will increase: usually, the soil has higher albedo than vegetation. Thus, more solar radiation will be reflected and the surface will become cooler, and so will become the lower atmospheric layers which are in contact with the surface. The atmospheric column will thus become more stable, and less convection will ensue. Since convection is one of the basic atmospheric processes leading to precipitation, the reduction of convection will imply a reduction in precipitation, with a further negative effect on vegetation. A simple model of this mechanism indicates that two possible stable states emerge, corresponding respectively to a vegetated state and to bare soil conditions [4].

Of course, things are more complicated than this simple example: in addition to the vegetation-albedo feedback, a reduction in vegetation implies a lower moisture (and latent heat) flux from the surface to the atmosphere, a fact which further contributes to the stabilization of the atmospheric column [9]. In many instances, the reduction in latent heat flux is probably more important than the albedo change [1]. The reduction in precipitation, then, is often associated with a change in the pressure distribution over large areas, with a corresponding modification in the regional atmospheric circulation and monsoon dynamics [6].

In such a complex system, it is not easy to foresee what will be the response to a perturbation in the external forcings: some feedback mechanisms can amplify in unexpected ways the original disturbance, while others can reduce its effects. To try to disentangle the workings of the climate system, we need to develop mathematical descriptions, or models, of climate and its components.

2 The mathematics of climate

Science is based on data, observations, and experiments, and climate science is no exception. In the last thirty years, our understanding of climate has greatly advanced, mainly thanks to extended measurement networks which include ground data, atmospheric soundings, oceanic measurements and satellite observations. Data are continuously collected, and – in many instances – they are made available to the whole scientific community and to citizens following an "open data" view. Data, however, need to be managed, analyzed and used to develop, validate and falsify theories and models. And as often happens in science, models and theories are framed in a mathematical language.

To build a climate model, we are forced to simplify the problem and decide what is important and should be kept in the model, and what could be, at least temporarily, discarded. This distinction is by no means universal, and the choice depends on the type of problem we want to tackle. In principle, all processes are important, and we should describe the dynamics of individual thunderstorms together with the slow variations of the Earth's orbital parameter. Such an approach is unfeasible and probably not even useful: instead of a brute-force approach where we put everything in the same pot, it is more interesting to develop a hierarchy of models and descriptions, trying to address specific problems and understand what are the causes of a given behaviour.

Traditionally, climate models have been divided in a sort of hierarchy, from the simplest to the most complex, with different purposes and different types of simplifications [22].

Process (or conceptual) models, based on explicit formulations and simplified descriptions, aim at addressing specific questions and feedback mechanisms, such as the Charney mechanism described above or the ice-albedo feedback [15]. Among these models, the so-called Energy Balance Models have a long history [21], and can be used to study issues such as paleoclimatic variability on long time scales or extreme climatic conditions [14] and, more recently, the climate of planets outside the Solar System [27, 30, 32]. These models usually consider only one spatial dimension along latitude, or even reduce the whole planet to a single homogeneous element (box) with no internal spatial dependence.

Alternatively, one can develop simplified models for the vertical structure of the atmosphere, the so-called radiative-convective models [29], to study processes associated with absorption and re-emission of infrared radiation in the atmospheric column. Other models try to describe the multiple states of the thermohaline circulation [28], the alternation between glacials and interglacials [16, 23], the interaction between the atmosphere and the ocean in the ENSO phenomenon [18], and the interaction between the atmosphere and the vegetation [8]. All these models are amenable to serious mathematical analysis, and often are formidable tools to learn – and teach – the fundamental processes of climate dynamics. Of course, they have an extremely limited ability to produce detailed climate predictions, and should be intended mainly as research and understanding tools.

At the other end of the spectrum, Global Climate Models (GCMs) are based on our current best knowledge of the climate system [31]. Such models try to include "all that we know" about climate, and aim at producing reliable climate projections for future decades. A lot of effort and energy has been spent in order to build, validate and improve GCMs, and the last thirty years have witnessed a significant progress in our ability to realistically model climate, its variability and its response to changes in the external forcing. Such models are usually extremely heavy, require large computing power and enormous storage to be run, and – a somehow disturbing aspect – are so complicated that usually no single person knows the whole model: each research group works on a specific aspect, and the final model is built by patching together the different components. In the last years, some of the big GCM source codes have become publicly available, and the model themselves are to be considered as community models. This is the case, for example, for the "EC-Earth" Earth System Model [17], which is maintained by a European Consortium which includes several countries and more than twenty research institutions (see http://ecearth.knmi.nl/).

Of course, nothing is perfect under the sky, and Global Climate Models are no exception. For example, their spatial resolution is usually of the order of 50-100 km, and all processes below this scale cannot be directly resolved. Thunderstorms, surface inhomogeneities, steep orography, cloud physics, and precipitation intermittency, just to name a few, are not directly seen by the GCMs. Instead, these processes are "parameterized": their small-scale dynamics is implicitly described in terms of the large-scale properties, that is, the small-scale (fast) variability is slaved to the large-scale (slow) modes. Sometimes this procedure works well, sometimes it does not, and it is not easy to know in advance when it provides acceptable results. Here, we enter the realm of turbulence, turbulent convection, land-vegetation-atmosphere interaction and microphysics, and there is room for a lot of work on fundamental aspects of process interaction across multiple scales and on the development of more motivated parameterization schemes [26]. And, even, for addressing the question of whether parameterization is always possible.

Besides the issue of parameterization, GCMs also have to cope with poorly understood phenomena, which are described by empirical laws: while the equations for the large-scale motions in the atmosphere and the ocean are reasonably known, our current description of vegetation dynamics is rather coarse, and many basic processes in land-atmosphere interaction are treated only empirically.

Somewhere in between process models and GCMs, one finds the so-called Earth System Models of Intermediate Complexity (EMICs), which are a sort of stripped-down version of Global Climate Models, with simplified parameterizations, and can be run on local workstations [7]. One notable example is the Planet Simulator developed at the University of Hamburg (http://www.mi.uni-hamburg.de/index.php?id=216), which is used to study the climate of the Earth, Mars, and Titan, and to explore different aspects of climate-biosphere interactions [10, 13]. EMICs are complicated, but they can be managed and understood by individual researchers or by small research groups. Their predictive ability is probably more limited than that of the big GCMs, but they can describe several processes at once and can be used to explore a large variety of climatic conditions. Depending upon the point of view, one can consider EMICs as a more complicated version of conceptual models, and use them to explore process interactions and extreme climatic conditions. Alternatively, one can see EMICs as simplified GCMs which run faster and require less computing power, and use them to obtain long paleoclimatic simulations or an ensemble of many different realizations of climate variability under a specified external forcing. Also EMICs have several drawbacks, but they are probably the best ground where climate scientists, fluid dynamicists and mathematicians can interact with reciprocal satisfaction, as illustrated, for example, by many interesting results on the bifurcation analysis of oceanic flows [11].

3 Towards a theory of climate

In the last thirty years, we have learned a lot about climate, thanks mainly to new and extensive measurements and to the efforts spent in developing detailed descriptions of many climatic processes. As a result, the scientific community has now developed powerful modeling tools which include process models, EMICs, and Global and Regional Climate Models. The progress and success of climate modeling is huge, and we are now able to produce reliable projections for the climate of the coming decades, at least on continental and possibly regional scales. What clearly emerges is that the continuous emission of greenhouse gases is leading to a further temperature increase, continuing and perhaps emphasizing the warming trend which is evident from the data of the last 150 years. Thus, both mitigation and adaptation policies are necessary, and they should be discussed and evaluated in front of cost-benefit balances without hesitation.

However, it would be a capital mistake to think that we have had enough science, and now we only need to run the models on larger and larger computers to know what will happen in the coming decades. Climate still has many aspects which are poorly understood, including the role of small-scale processes and the whole issue of cross-scale interactions, the dynamics of clouds and of convection, the direct and indirect effects of aerosols, the role of biospheric processes and many aspects of ocean-atmosphere exchanges. To address these issues, the whole hierarchy of modeling tools is necessary, as no single approach is a priori preferable, and new ideas and interpretations should be developed. The hydrological cycle, for example, is one of the most important components of the climate of our planet, and it has a crucial impact on our own life. Still, precipitation intensity and variability are poorly reproduced by climate models, and a huge effort on further investigating such themes is required.

While continuing the necessary and fundamental efforts on data collection, storage and analysis, and the development of more sophisticated modeling tools, we also need to come up with a theory of climate. Such theory should try to put together the different pieces of the climatic puzzle, addressing the most important open questions, developing the proper mathematical descriptions, in a world-wide initiative to understand one of the most fascinating and important manifestations of Planet Earth.

Acknowledgements The author's research activity on climate is largely supported by the Project of Interest "NextData" of the Italian Ministry for Education, University and Research. Many thanks to Maria for reading the manuscript and providing insightful comments.

References

1. Baudena, M., D'Andrea, F., Provenzale, A.: A model for soil-vegetation-atmosphere interactions in water-limited ecosystems. Water Resour. Res. **44**, 1–9 (2008)
2. Benzi, R., Parisi, G., Sutera, A., Vulpiani, A.: Stochastic resonance in climatic change. Tellus, **34**, 10–16 (1982)
3. Bordi, I., Fraedrich, K., Sutera, A., Zhu, X.: Transient response to well-mixed greenhouse gas changes. Theor. Appl. Climatol., **109**, 245–252 (2012)
4. Brovkin, V., Claussen, M., Petoukhov, V., Ganopolski, A.: On the stability of the atmosphere-vegetation system in the Sahara/Sahel region. J. Geophys. Res. **103**(31), 613–624 (1998)
5. Charney, J.: Dynamics of deserts and droughts in the Sahel. Q. J. Roy Meteor. Soc. **101**, 193–202 (1975)
6. Claussen, M.: On multiple solutions of the atmosphere–vegetation system in present-day climate. Global Change Biology **4**, 549–559 (1998)
7. Claussen, M. and 18 others: Earth system models of intermediate complexity: closing the gap in the spectrum of climate system models. Climate Dynamics **18**, 579–586 (2002)
8. Cresto Aleina, F., Baudena, M., D'Andrea, F., Provenzale, A.: Multiple equilibria on planet Dune: climate-vegetation dynamics on a sandy planet. Tellus B (2013). doi: 10.3402/tellusb.v65i0.17662

9. D'Andrea, F., Provenzale, A., Vautard, R., De Noblet-Decoudré, N.: Hot and cool summers: Multiple equilibria of the continental water cycle. Geophys. Res. Lett. (2006). doi: 10.1029/2006GL027972
10. Dekker, S.C., de Boer, H.J., Brovkin, V., Fraedrich, K., Wassen, M.J., Rietkerk, M.: Biogeophysical feedbacks trigger shifts in the modelled climate system at multiple scales. Biogeosciences **7**, 1237–1245 (2010)
11. Dijkstra, H.A.: Nonlinear Physical Oceanography: A Dynamical Systems Approach to the Large Scale Ocean Circulation and El Nino. Springer, New York, USA (2005)
12. Dijkstra, H.A.: Nonlinear Climate Dynamics. Cambridge University Press, Cambridge, UK (2013)
13. Fraedrich, K., Jansen, H., Kirk, E., Luksch, U., Lunkeit, F.: The Planet Simulator: Towards a user friendly model. Meteorologische Zeitschrift **14**, 299–304 (2005)
14. Ghil, M.: Cryothermodynamics: the chaotic dynamics of paleoclimate. Physica D **77**, 130–159 (1994)
15. Ghil, M., Childress, S.: Topics in Geophysical Fluid Dynamics: Atmospheric Dynamics, Dynamo Theory and Climate Dynamics. Springer, New York, USA (1987)
16. Gildor, H., Tziperman, E.: A sea ice climate switch mechanism for the 100-kyr glacial cycles. J. Geophysical Res. Oceans **106**, 9117–9133 (2001)
17. Hazeleger, W., Bintanja, R.: Studies with the EC-Earth seamless earth system prediction model. Climate Dynamics **39**, 2609–2610 (2012)
18. Jin, F.F., Neelin, J.D., Ghil, M.: El Nino on the Devil's Staircase: Annual subharmonic steps to chaos. Science **264**, 70–72 (1994)
19. Lovelock, J.E.: Gaia. A new look at life on Earth. Oxford University Press, Oxford, UK (1979)
20. Milankovitch, M.: Canon of Insolation and the Ice Age Problem. Royal Serbian Academy, Belgrade (1941)
21. North, G.R., Cahalan, R.F., Coakley, J.A.Jr: Energy Balance Climate Models. Rev. Geophysics and Space Physics **19**, 91–121 (1981)
22. McGuffie, K., Henderson-Sellers, A.: A Climate Modelling Primer. Wiley, Chichester, UK (2005)
23. Paillard, D.: Glacial cycles: Toward a new paradigm. Reviews of Geophysics **39**, 325–346 (2001)
24. Peixoto, J.P., Oort, A.H.: Physics of Climate. AIP Press, New York, USA (1992)
25. Pierrehumbert, R.T.: Principles of Planetary Climate. Cambridge University Press, Cambridge, UK (2010)
26. Rietkerk, M. and 15 others: Local ecosystem feedbacks and critical transitions in the climate. Ecological Complexity **8**, 223–228 (2011)
27. Spiegel, D.S., Menou, K., Scharf, C.A.: Habitable Climates. The Astrophysical Journal **681**, 1609–1623 (2008)
28. Stommel, H.: Thermohaline Convection with Two Stable Regimes of Flow. Tellus **13**, 224-230 (1961)
29. Tyler, D.R., Catling, D.C.: An analytic radiative-convective model for planetary atmospheres. The Astrophysical Journal (2012). doi:10.1088/0004-637X/757/1/104
30. Vladilo, G., Murante, G., Silva, L., Provenzale, A., Ferri, G., Ragazzini, G.: The habitable zone of Earth-like planets with different levels of atmospheric pressure. The Astrophysical Journal (2013). doi:10.1088/0004-637X/767/1/65
31. Washington, W.M., Parkinson, C.L.: An Introduction to Three-Dimensional Climate Modeling. University Science Books, Sausalito, CA, USA (2005)
32. Williams, D.M., Kasting, J.F.: Habitable Planets with High Obliquities. ICARUS **129**, 254–267 (1997)

Periodic Orbits of the N-body Problem with the Symmetry of Platonic Polyhedra

Giovanni Federico Gronchi

Abstract We review some recently discovered periodic orbits of the N-body problem [8], whose existence is proved by means of variational methods. These orbits are minimizers of the Lagrangian action functional in a set of T-periodic loops, equivariant for the action of a group G and satisfying some topological constraints. Both the group action and the topological constraints are defined using the symmetry of Platonic polyhedra.

1 Introduction

The N-body problem is the study of the motion of N particles, regarded as point masses, which are subject to their mutual gravitational interaction. This problem has been investigated for a long time and is useful for different purposes, e.g. to study the stability of the solar system planets, or to predict possible collisions of asteroids with the Earth.

For $N \geq 3$ the problem is not integrable (in the sense of classical Mechanics) and, even worse, chaoticity phenomena may prevent a reliable computation of the solutions with numerical methods over the desired time span.

The importance of periodic orbits in the study of the N-body problem is well expressed by the words of Poincaré: *ce qui rend ces solutions périodiques aussi précieuses, c'est qu'elle sont, pour ainsi dire, la seule brèche par où nous puissions pénétrer dans une place jusqu'ici réputée inabordable* [13].

In the literature there are different ways to prove the existence of periodic solutions of this problem: continuation of known solutions, expansions by series, topological methods, etc. Here we consider solutions whose existence is proved by means of

G.F. Gronchi (✉)
Dipartimento di Matematica, Università di Pisa, Largo B. Pontecorvo 5, 56127 Pisa, Italy
e-mail: gronchi@dm.unipi.it

A. Celletti, U. Locatelli, T. Ruggeri and E. Strickland (eds.): *Mathematical Models and Methods for Planet Earth.* Springer INdAM Series 6, DOI 10.1007/978-3-319-02657-2_12,

Calculus of Variations. In the last years some new and unexpected periodic orbits of the N-body problem have been found by variational methods; one of these is the 'figure eight' [4,14], where three equal masses follow the same planar, eight-shaped curve, with the same time law and a shift of $1/3$ of the period one after the other. A motion of this kind is called *choreography* [17]. For other results on periodic orbits of the N-body problem see [2,7,8] and references therein.
In this short paper we review some of the results in [8], where the symmetry groups of Platonic polyhedra play a fundamental role.

2 Calculus of variations for periodic orbits

We can choose the units so that the equations of motion of the N-body problem are

$$m_i \ddot{u}_i = U_{u_i}(u), \quad i = 1, \ldots, N, \tag{1}$$

where $m_1, \ldots, m_N$ are the masses of the particles and

$$U(u) = \sum_{1 \le h < k \le N} \frac{m_h m_k}{|u_h - u_k|}$$

is the potential. The map $\mathbb{R} \ni t \to u = (u_1, \ldots, u_N) \in \mathbb{R}^{3N}$ describes the evolution of the positions of the N bodies in $\mathbb{R}^3$.
We introduce the kinetic energy

$$K(\dot{u}) = \frac{1}{2} \sum_{i=1}^{N} m_i |\dot{u}_i|^2.$$

For a fixed period $T > 0$ we can give a variational formulation to the search for periodic orbits of (1): the solutions are stationary points of the Lagrangian action functional

$$\mathscr{A}(u) = \int_0^T K(\dot{u}(t)) + U(u(t))\, dt \tag{2}$$

on a set of admissible curves. In particular, we can search for periodic solutions which are minimum points of $\mathscr{A}$.
A natural environment for this problem is the Sobolev space $H^1_T(\mathbb{R}, \mathbb{R}^{3N})$ of T-periodic *loops* (i.e. closed curves) in H^1, with norm

$$\|u\|_{H^1_T} = \left(\int_0^T |u(t)|^2 dt + \int_0^T |\dot{u}(t)|^2 dt \right)^{1/2},$$

where $\dot{u} \in L^2$ is the weak derivative of u in H^1.
Using the integrals of the center of mass we can assume $\sum_{h=1}^{N} m_h u_h = 0$. We introduce the configuration space

$$\mathscr{X} = \left\{ x = (x_1, \ldots, x_N) \in \mathbb{R}^{3N} : \sum_{h=1}^{N} m_h x_h = 0 \right\}$$

and consider the loop space $\Lambda = H^1_T(\mathbb{R}, \mathscr{X})$.

Let $E \subseteq \Lambda$ be a set of loops. We say that a functional $\mathscr{J} : \Lambda \to \mathbb{R}$ is *coercive* on E if $\lim_{k\to\infty} \mathscr{J}(u^{(k)}) = +\infty$ for each sequence $\{u^{(k)}\}_{k\in\mathbb{N}} \subset E$ such that $\lim_{k\to\infty} \|u^{(k)}\|_{H^1_T} = +\infty$. Standard methods of Calculus of Variations, going back to Tonelli [18], ensure the existence of minimum points of $\mathscr{J}$ in $\overline{E}$ (the closure of E in Λ) provided $\mathscr{J}$ is coercive on E.
However, the action functional $\mathscr{A}$ is not coercive on the whole space Λ; we can even find sequences $\{u^{(k)}\}_{k\in\mathbb{N}} \subset \Lambda$ such that

$$\lim_{k\to\infty} \|u^{(k)}\|_{H^1_T} = +\infty \qquad \text{and} \qquad \lim_{k\to\infty} \mathscr{A}(u^{(k)}) = 0.$$

Another obstruction is that minimizers of $\mathscr{A}$ may have collisions: this is a consequence of Sundman's estimates [1,20], and had already been noted by Poincaré [16] for the case of three bodies. In fact, if u_* is a minimizer with collisions, then close to a collision time t_c we have

$$|u_*(t)| = O(|t_c - t|^{2/3}),$$

therefore the contribution of collisions to the action $\mathscr{A}$ is finite.

We can recover coercivity by introducing constraints to the admissible loops. For example in [10] the author proves that the whole family of isochronous Keplerian orbits, including degenerate collision-ejection solutions, minimizes the Lagrangian action of the Kepler problem in the set of T–periodic loops in $H^1(\mathbb{R},\mathbb{R}^2)$ winding around the center of attraction exactly once, either clockwise or counter–clockwise. This is a *topological constraint* and is sufficient to have coercivity.

Another way to obtain coercivity is by *symmetry constraints*. For example in [5, 6] the authors restrict the minimization of the action to the maps $u(t) \in \Lambda$ such that:

$$u(t+T/2) = -u(t).$$

In both these examples the trajectories of any sequence $\{u^{(k)}\}_{k\in\mathbb{N}}$ with $\lim_{k\to\infty} \|u^{(k)}\|_{H^1_T} = +\infty$ is such that

$$\lim_{k\to\infty} \max_{s,t\in\mathbb{R}} |u^{(k)}(s) - u^{(k)}(t)| = +\infty,$$

so that the kinetic part of the action diverges:

$$\lim_{k\to\infty} \int_0^T K(\dot{u}^{(k)}(t))\, dt = +\infty.$$

A general formulation for the use of symmetry constraints to prove the existence of periodic orbits is given in [7]. Here the authors restrict the admissible curves to the set Λ_G of loops equivariant with respect to the action of a finite group G on the space $\Lambda = H^1_T(\mathbb{R}, \mathscr{X})$.

Assume $m_i = m_j$ for $1 \le i < j \le N$. Consider a finite group G and its three representations

$$\tau : G \to O(2), \qquad \rho : G \to O(3), \qquad \sigma : G \to \Sigma_N,$$

that are homomorphisms on the orthogonal groups $O(2)$, $O(3)$ and on the symmetric group Σ_N of permutations of N elements. Through τ, ρ, σ we can define an action of G on the loop space Λ by

$$G \times \Lambda \ni (g,u) \mapsto g \cdot u \in \Lambda, \qquad (g \cdot u)_j(t) := \rho(g) u_{\sigma(g^{-1})(j)}(\tau(g^{-1})t)$$

and an action on the configuration space $\mathscr{X}$ by

$$G \times \mathscr{X} \ni (g,x) \mapsto g \cdot x \in \mathscr{X}, \qquad (g \cdot x)_j := \rho(g) x_{\sigma(g^{-1})(j)},$$

for each $j = 1 \ldots N$, $t \in \mathbb{R}$. We denote by Λ_G the set of G-equivariant loops in Λ: $u \in \Lambda$ belongs to Λ_G if and only if

$$g \cdot u(t) = u(t), \qquad \forall g \in G, \forall t \in \mathbb{R}.$$

Moreover, let $\mathscr{X}^G$ be the set of configurations in $\mathscr{X}$ which are fixed by every element of G. A necessary and sufficient condition for coercivity (see [7]) is given by

Proposition 1 *The Lagrangian action $\mathscr{A}$, restricted to the set of G-equivariant loops Λ_G, is coercive if and only if $\mathscr{X}^G = \{0\}$.*

Since $\mathscr{X}^G = \{0\}$ if and only if

$$\frac{1}{T} \int_0^T u_j(t)\, dt = 0, \qquad j = 1, \ldots, N, \tag{3}$$

the trajectories of the N particles must share a common center (in the sense of the integral average). This is a strong constraint to the motion.

We discuss now the problem of collisions. In the literature there are different methods to show that a minimizer of $\mathscr{A}$ is *collision-free*. We consider two classes of methods:

i) **Level estimates:** compute an *a priori* lower bound a for $\mathscr{A}(u_*)$, where u_* is any minimizer with collisions, and find an admissible loop v such that

$$\mathscr{A}(v) < a \le \mathscr{A}(u_*).$$

ii) **Local perturbations:** for each minimizer u_* with collisions find a small admissible perturbation $u_* + \xi$ such that

$$\mathscr{A}(u_* + \xi) < \mathscr{A}(u_*).$$

We recall two results that are useful to define local perturbations. The first is due to C. Marchal [3]. In the Kepler problem with center of force O the action of the parabolic collision-ejection arc AOB (going from A to B through O) is greater than the action of the isochronous direct Keplerian arc joining A with B and, provided $A \neq B$, it is also greater than the action of the indirect Keplerian arc. If $A = B$ the indirect arc does not exist. We can construct local perturbations of collision solutions using these Keplerian arcs combined with a blow–up technique (see [7]).

The second result is also based on an idea by Marchal [3, 12], and has been later generalized in [7]. Consider a family of admissible perturbations u^σ, parametrized by the directions σ spanning the whole unit sphere $\mathbb{S}^2$. If the integral average $\frac{1}{4\pi}\int_{\mathbb{S}^2} \mathscr{A}(u^\sigma) d\sigma$ of the values of the action over $\mathbb{S}^2$ is less than $\mathscr{A}(u_*)$, then there exists a perturbation with lower action than u_*, even if we cannot exhibit it explicitly.

3 Platonic polyhedra and orbits with symmetry and topological constraints

We present some of the results in [8], where new periodic orbits of the N-body problem with the symmetry of Platonic polyhedra have been introduced.
The symmetry of Platonic polyhedra was already associated to the motion of celestial bodies by J. Kepler at the end of the XVI century [11].

Let $\mathcal{R} \in \{\mathcal{T}, \mathcal{O}, \mathcal{I}\}$, where $\mathcal{T}, \mathcal{O}, \mathcal{I}$ are the rotation groups of tetrahedron, octahedron, icosahedron. Hexahedron and dodecahedron have the same rotation group as octahedron and icosahedron respectively. We set $N = |\mathcal{R}|$ and assume $m_i = m_j = 1$ for $i, j = 1, \dots, N$.
We restrict the Lagrangian action to open cones

$$\mathcal{K} = \{u \in H^1_T(\mathbb{R}, \mathcal{X}) : u \text{ satisfies (a), (b)}\}$$

where:

(a) The motion u_j of the j-th particle is determined by the motion u_1 of the first particle (that we call *generating particle*) through the relation

$$u_j = R_j u_1, \qquad j \in \{1, \dots, N\}$$

with $\{1, \dots, N\} \ni j \to R_j \in \mathcal{R}$ a bijection such that $R_1 = I$;

(b) The motion u_1 of the generating particle belongs to a given free homotopy class in $\mathbb{R}^3 \setminus \Gamma$, where

$$\Gamma = \cup_{R \in \mathcal{R} \setminus \{I\}} r(R)$$

with $r(R)$ the rotation axis of R.

Conditions (a), (b) are a symmetry and a topological constraint respectively. The superposition of these constraints allows us to produce orbits that do not share a common center, unlike the ones in [7]. Moreover, assuming condition (a) holds, the set Γ in condition (b) is exactly the set of points where collisions take place, if any.

3.1 Encoding the cones $\mathcal{K}$

We describe two ways to encode the topological constraints defining the cones $\mathcal{K}$. Let $\tilde{\mathcal{R}}$ be the full symmetry group (including reflections) related to $\mathcal{R}$. The reflection planes induce a tessellation of the unit sphere $\mathbb{S}^2$, as in Figure 1, with $2N$ spherical triangles. Each vertex of such triangles corresponds to a pole $p \in \mathcal{P} = \Gamma \cap \mathbb{S}^2$. Select one of these triangles, say τ. By a suitable choice of a point $q \in \partial\tau$ (see Fig. 1) we can define an Archimedean polyhedron $\mathcal{Q}_{\mathcal{R}}$, which is the convex hull of the orbit of q under $\mathcal{R}$, and therefore is strictly related with the symmetry group $\mathcal{R}$. For details see [8].

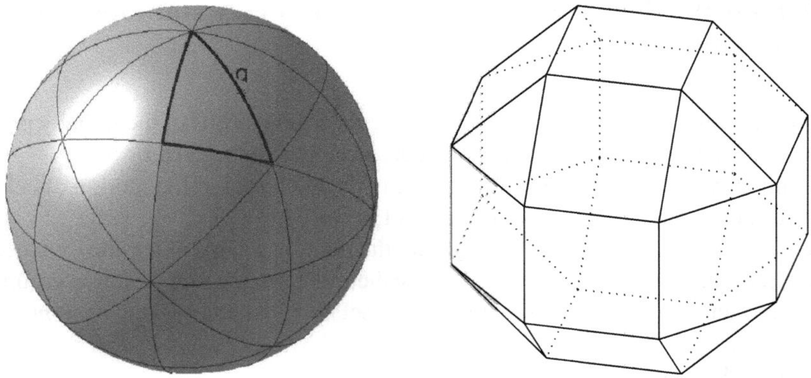

Fig. 1 Tessellation of $\mathbb{S}^2$ for $\mathscr{R} = \mathscr{O}$ and the Archimedean polyhedron $\mathscr{Q}_{\mathscr{O}}$

We can characterize a cone $\mathscr{K}$ by a periodic sequence $\mathrm{t} = \{\tau_k\}_{k\in\mathbb{Z}}$ of triangles of the tessellation such that τ_{k+1} shares an edge with τ_k and $\tau_{k+1} \neq \tau_{k-1}$ for each $k \in \mathbb{Z}$. This sequence is uniquely determined by $\mathscr{K}$ up to translations, and describes the homotopy class of the admissible paths followed by the generating particle (see Fig. 2, left).

We can also characterize $\mathscr{K}$ by a periodic sequence $v = \{v_k\}_{k\in\mathbb{Z}}$ of vertexes of $\mathscr{Q}_{\mathscr{R}}$ such that the segment $[v_k, v_{k+1}]$ is an edge of $\mathscr{Q}_{\mathscr{R}}$ and $v_{k+1} \neq v_{k-1}$ for each $k \in \mathbb{Z}$. Also the sequence v is uniquely determined by $\mathscr{K}$ up to translations, and with it we can construct a piecewise linear loop, joining consecutive vertexes v_k with constant speed, that represents a possible motion of the generating particle (see Fig. 2, right).

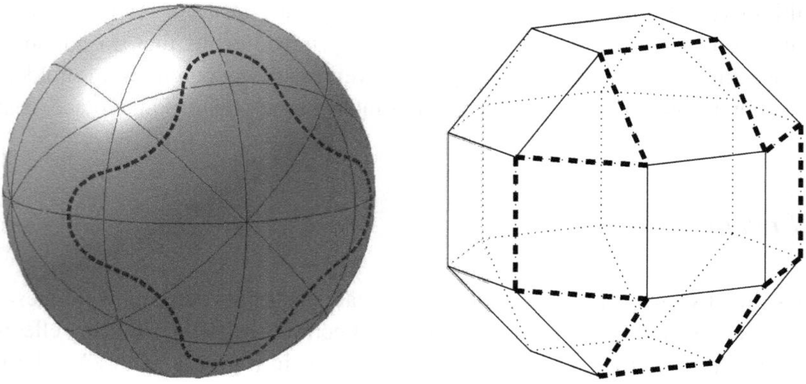

Fig. 2 Encoding a cone $\mathscr{K}$. Left: the dashed path on $\mathbb{S}^2$ describes the periodic sequence t of triangles of the tessellation. Right: the dashed piecewise linear path describes the corresponding periodic sequence v of vertexes of $\mathscr{Q}_{\mathscr{O}}$

3.2 Existence of minimizers

We consider cones $\mathscr{K} \subset \Lambda$ such that the trajectory of u_1 does not wind around one rotation axis only. In this case we can prove that $\mathscr{A}|_{\mathscr{K}}$ is coercive (see [8], Proposition 4.1).

If $\mathscr{A}$ is coercive, standard methods of Calculus of Variations yield the existence of a minimizer. Let $\{u^{(h)}\}_{h\in\mathbb{N}} \subset \mathscr{K}$ be a minimizing sequence, i.e. $\lim_{h\to\infty} \mathscr{A}(u^{(h)}) = \inf_{u\in\Lambda} \mathscr{A}(u)$. Up to subsequences, we can find a constant $M > 0$ such that $\mathscr{A}(u^{(h)}) \leq M$ for each $h \in \mathbb{N}$. We can take M equal to the action of the piecewise linear loop defined by the sequence ν that characterizes $\mathscr{K}$. Then coercivity implies

$$\|u^{(h)}\|_{H^1_T} \leq C, \qquad h \in \mathbb{N} \tag{4}$$

for a constant $C > 0$. Since $\Lambda = H^1_T(\mathbb{R}, \mathscr{X})$ is a reflexive space, there exists a subsequence of $\{u^{(h)}\}_{h\in\mathbb{N}}$ weakly converging to a loop u_* in Λ. From the bound (4) we also have

$$\|u^h\|_\infty \leq C, \qquad |u^{(h)}(t_1) - u^{(h)}(t_2)| \leq C|t_1 - t_2|^{1/2},$$

therefore, by the Ascoli-Arzelà theorem, there exists a subsequence of $\{u^{(h)}\}_{h\in\mathbb{N}}$ uniformly converging to u_* on compact sets. Hence $u_* \in \Lambda \cap \overline{\mathscr{K}}$, with $\overline{\mathscr{K}}$ the closure of $\mathscr{K}$ in the C^0 topology.
Moreover, the action functional $\mathscr{A}$ is weakly lower semi-continuous in Λ. In fact, if $\{u^{(h)}\}_{h\in\mathbb{N}}$ weakly converges to u_* in Λ, then it converges to u_* also uniformly (up to subsequences). By Cauchy-Schwartz inequality we have

$$\liminf_{h\to\infty} \|u^{(h)}\|^2_{H^1_T} \geq \|u_*\|^2_{H^1_T}. \tag{5}$$

Then relation (5) and the uniform convergence of $\{u^{(h)}\}_{h\in\mathbb{N}}$ imply

$$\liminf_{h\to\infty} \int_0^T |\dot{u}^{(h)}(t)|^2\, dt \geq \int_0^T |\dot{u}_*(t)|^2\, dt,$$

and

$$\lim_{h\to\infty} \int_0^T U(u^{(h)}(t))\, dt = \int_0^T U(u_*(t))\, dt,$$

so that

$$\liminf_{h\to\infty} \mathscr{A}(u^{(h)}) \geq \mathscr{A}(u_*).$$

If $\{u^{(h)}\}_{h\in\mathbb{N}}$ is a minimizing sequence we conclude that the limit u_* is a minimum point of $\mathscr{A}$.

There are infinitely many cones $\mathscr{K}$ and, therefore, infinitely many minimizers $u_* \in \overline{\mathscr{K}}$ of $\mathscr{A}|_{\mathscr{K}}$. However, we are interested only in classical solutions, i.e. in collision-free minimizers. Next section is devoted to the exclusion of collisions for minimizers in some particular classes of cones $\mathscr{K}$.

3.3 Collisions

Let $\mathfrak{S}$ be the set of loops with collisions. Since $\partial\mathscr{K} \subset \mathfrak{S}$, a minimizer u_* has a collision at $t = t_c$ if and only if $u_{*,1}(t_c) \in \Gamma$.
Due to the topological constraints, the proof that a minimizer u_* is collision-free cannot use Marchal's idea of averaging the action on a sphere [12], nor of averaging over a circle as in [7]. We shall exclude total collisions by level estimates, and partial collisions by local perturbations and by a uniqueness result for solutions of a class of differential equations with singular data.

3.3.1 Total collisions

For some cone $\mathscr{K}$ there exist $R \in \mathscr{R}$ and $M > 0$ such that the additional symmetry

$$u_1(t+T/M) = Ru_1(t), \tag{6}$$

is compatible with membership to $\mathscr{K}$. We can restrict again the minimization to the loops in $\mathscr{K}$ that fulfill (6). In this case, if the minimizer u_* has a total collision, then it has M total collisions per period.
We can estimate the action of u_* with the action of a homographic collision-ejection motion, with a minimal central configuration (see Propositions 5.1, 5.4 in [8]). In this way, if u_* has M collisions per period, we obtain the *a priori* estimate

$$\mathscr{A}(u_*) \geq a_{\mathscr{R},M} := 3\Big(\frac{N}{2}\Big)^{1/3}(\pi M U_0)^{2/3}T^{1/3},$$

where

$$U_0 = \min_{\rho(u)=1} U(u), \qquad \rho(u) = \frac{1}{\sqrt{N}}\Big(\sum_{h=1}^{N}|u_h|^2\Big)^{1/2}.$$

Rounded down values of $a_{\mathscr{R},M}$ for $T = 1$ and for different choices of $\mathscr{R}$, M are given in Table 1.

Table 1 Lower bounds $a_{\mathscr{R},M}$ for loops with M total collisions ($T = 1$)

$\mathscr{R} \backslash M$	1	2	3	4	5
$\mathscr{T}$	132.695	210.640	276.017	/	/
$\mathscr{O}$	457.184	725.734	950.981	1152.032	/
$\mathscr{I}$	2296.892	3646.089	4777.728	/	6716.154

For some sequences v, the action of the related linear piecewise loop v is lower than $a_{\mathscr{R},M}$. Therefore, minimizing the action $\mathscr{A}$ over the cones $\mathscr{K}$ defined by those sequences v, with the additional symmetry (6), yields minimizers u_* without total collisions.

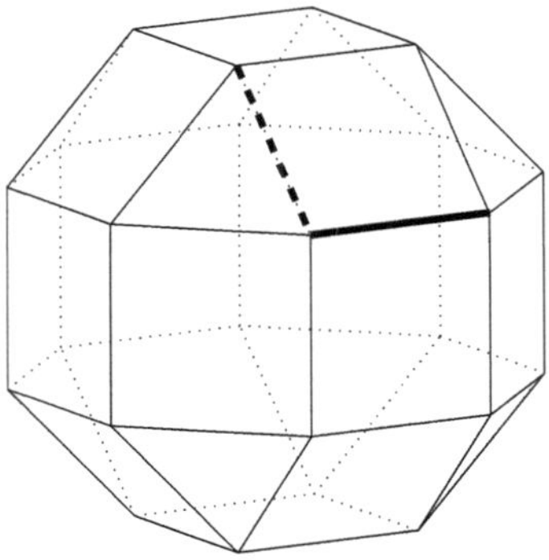

Fig. 3 For $\mathscr{R} = \mathscr{O}$ two sides of different kind are enhanced on the Archimedean polyhedron $\mathscr{Q}_{\mathscr{O}}$. The dashed side joins a triangle and a square; the solid side joins two squares

The integrals appearing in this computation can be expressed by means of elementary functions. Indeed we have to compute a sum of integrals of (at most) two kinds due to the structure of the potential U and to the symmetry of the Archimedean polyhedron $\mathscr{Q}_{\mathscr{R}}$, which has at most two sides of different kind[1], see Figure 3. We obtain the following formula (see [8], Proposition 5.5) for the action of the loop v:

$$\mathscr{A}(v) = \frac{3}{2 \cdot 4^{1/3}} N \ell^{2/3} (n_1 \upsilon_1 + n_2 \upsilon_2)^{2/3} T^{1/3}, \tag{7}$$

where $n_i, i = 1,2$ are the numbers of sides of the two different kinds in the trajectory of v, ℓ is the length of the sides (assuming the polyhedra are inscribed in the unit sphere $\mathbb{S}^2$) and $\upsilon_i, i = 1,2$ are the values of explicitly computable integrals, see Table 2.

Table 2 Numerical values of ℓ, υ_1, υ_2

$\mathscr{R}$	$\mathscr{T}$	$\mathscr{O}$	$\mathscr{I}$
ℓ	1.0	0.7149	0.4479
υ_1	9.5084	20.3225	53.9904
υ_2	9.5084	19.7400	52.5762

3.3.2 Excluding partial collisions

Assume u_* has a partial collision at time $t = t_c$. Because of the symmetry, the generating particle must collide on a rotation axis, say r. Actually all the particles collide in separate clusters, and the number of particles in each colliding cluster is the order $o_{\mathscr{C}}$ of the maximal cyclic subgroup $\mathscr{C}$ of rotations around r in $\mathscr{R}$. We can also assume that collisions are isolated in time [2,7].

[1] Indeed for $\mathscr{R} = \mathscr{T}$ all sides are of the same kind.

The motion of the generating particle $t \to u_{*,1}(t)$, colliding on r at $t = t_c$, satisfies an equation of the form

$$\ddot{w} = \alpha \frac{(R_\pi - I)w}{|(R_\pi - I)w|^3} + V_1(w), \qquad \alpha = \sum_{j=1}^{\mathfrak{o}_{\mathscr{C}}-1} \frac{1}{\sin\left(\frac{j\pi}{\mathfrak{o}_{\mathscr{C}}}\right)}, \tag{8}$$

where R_π is the rotation of π around the axis r and $V_1(w)$ is a smooth function.

Assume t_c=0 and let $w : (0,\bar{t}) \to \mathbb{R}^3$ be a maximal solution of (8) with $\lim_{t\to 0^+} w(t) = 0$. Then there exists $b \in \mathbb{R}$ and a unit vector n^+, with $\mathsf{n}^+ \perp r$, such that

$$\lim_{t\to 0^+} \frac{\dot{w}(t) + R_\pi \dot{w}(t)}{2} = b\mathsf{e}_r, \qquad \lim_{t\to 0^+} \frac{w(t)}{|w(t)|} = \mathsf{n}^+ . \tag{9}$$

The vector n^+ corresponds to the ejection limit direction. This singularity can be regarded asymptotically as a parabolic binary collision. In fact we can perform a blow-up [7], [19] by considering the rescaled functions

$$w^\lambda : [0,1] \to \mathbb{R}^3, \qquad w^\lambda(\tau) = \lambda^{2/3} w(\tau/\lambda),$$

with $\lambda > 1/\bar{t}$. The family of functions $\{w^\lambda\}_\lambda$ converges uniformly in $[0,1]$, as $\lambda \to +\infty$, to the parabolic ejection motion

$$s^\alpha(\tau)\mathsf{n}^+, \qquad s^\alpha(\tau) = (3^{3/2}/2)\alpha^{1/3}\tau^{1/3}.$$

Analogous statements hold assuming $\lim_{t\to 0^-} w(t) = 0$, with a unit vector n^-, $\mathsf{n}^- \perp r$ corresponding to the collision limit direction.

We say that a cone $\mathscr{K}$ is *simple* if the corresponding sequence $\mathfrak{t}$ does not contain a string $\tau_k \ldots \tau_{k+2\mathfrak{o}}$ such that

$$\bigcap_{j=0}^{2\mathfrak{o}} \overline{\tau_{k+j}} = p,$$

where $p \in \mathbb{S}^2$ is the pole of some rotation $R \in \mathscr{R} \setminus \{I\}$ and $\mathfrak{o}$ is the order of p.

Given a minimizing sequence $\{u^{(h)}\}_{h\in\mathbb{N}}$ we can assume that each loop $u_1^{(h)}$, describing the motion of the generating particle, when projected on $\mathbb{S}^2$, crosses the minimum number of triangles τ_k in a period (see [8], Proposition 4.4).
If the minimizer u_*, which is the limit of such a sequence, is such that $u_{*,1}$ has a collision on the axis r at time $t = t_c$, then we can define a *collision angle* θ that represents the angle between the incoming and outcoming asymptotic directions $\mathsf{n}^+, \mathsf{n}^-$, taking into account the complexity of trajectories in the minimizing sequence (this allows for example to distinguish between cases with $\theta = 0$ and $\theta = 2\pi$). Let $\mathfrak{o}_r$ be the order of the maximal cyclic group related to the collision axis r. If $\mathscr{K}$ is simple we have

$$-\frac{\pi}{\mathfrak{o}_r} \le \theta \le 2\pi.$$

If $\theta \neq 2\pi$ we can exclude partial collisions by local perturbations, that are constructed with direct or indirect arcs, and with the blow-up technique (see [8], Proposition 5.8). If $\theta = 2\pi$ we have:

i) $\mathsf{n}^+ = \mathsf{n}^-$.
ii) The plane $\pi_{r,\mathsf{n}}$ generated by r, $\mathsf{n} = \mathsf{n}^\pm$ is fixed by some reflection $\tilde{R} \in \tilde{\mathscr{R}}$.

In this case we cannot exclude the singularity by choosing between direct and indirect arc, because the indirect arc is not available. To exclude these kind of collisions we use the following uniqueness result (see [8], Proposition 5.9) for solutions of Equation (8) with singular initial data.

Proposition 2 *Let* $w_i : (0,\bar{t}_i) \to \mathbb{R}^3, \bar{t}_i > 0, i = 1,2$ *be two maximal solutions of* (8) *such that* $\lim_{t\to 0^+} w_i(t) = 0$. *If* h_i, b_i, n_i, $i = 1,2$ *are the corresponding values of the energy, and the values of* b *and* n^+ *given by* (9), *then*

$$h_1 = h_2\,, b_1 = b_2\,, \mathsf{n}_1 = \mathsf{n}_2 \qquad \textit{implies} \qquad \bar{t}_1 = \bar{t}_2\,, w_1 = w_2\,.$$

As a consequence, using the symmetry of the potential U, we have

Corollary 1 *Let* $w : (0,\bar{t}) \to \mathbb{R}^3$ *be a maximal ejection solution of* (8) *and* $\mathsf{n}^+ = \lim_{t\to 0^+} \frac{w(t)}{|w(t)|}$. *Assume the plane* π_{r,n^+} *generated by* r, n^+ *is fixed by some reflection* $\tilde{R}$ *in* $\tilde{\mathscr{R}}$. *Then*

$$w(t) \in \pi_{r,\mathsf{n}^+},\ \forall t \in (0,\bar{t})\,.$$

A similar result holds for collision solutions $w : (-\bar{t},0) \to \mathbb{R}^3$.

We can assume that all partial collisions of the generating particle satisfy conditions i), ii) above. Indeed different kinds of collisions are excluded by local perturbations. From Proposition 2 and Corollary 1 the generating particle must move on a reflection plane, between two rotation axes. This contradicts membership to $\overline{\mathscr{K}}$, except for particular topological constraints, defined by sequences ν winding around two axes only.
We conclude that, provided the cone $\mathscr{K}$ is simple, and the related sequence ν does not wind around one axis, nor two axes only, then minimizing the action $\mathscr{A}$ in $\mathscr{K}$ yields loops u_* free of partial collisions.

3.4 Main results

For $\mathscr{R} \in \{\mathscr{T},\mathscr{O},\mathscr{I}\}$ we can list a number of periodic sequences ν, determined by the elements $\nu_k, k = 0,\ldots,\kappa_\nu$ with κ_ν the minimal period of ν, such that there exists a T-periodic solution of the N-body problem in the same free homotopy class of ν (see [8], Theorem 4.1). For example, assuming $\mathscr{R} = \mathscr{O}$, we can consider

$$\nu \equiv [5,1,16,10,3,8,18,7,20,23,14,11,5],$$

where the numbering of vertexes of $\mathscr{Q}_\mathscr{O}$ is given in Figure 4. This sequence corresponds to the paths in Figure 2.

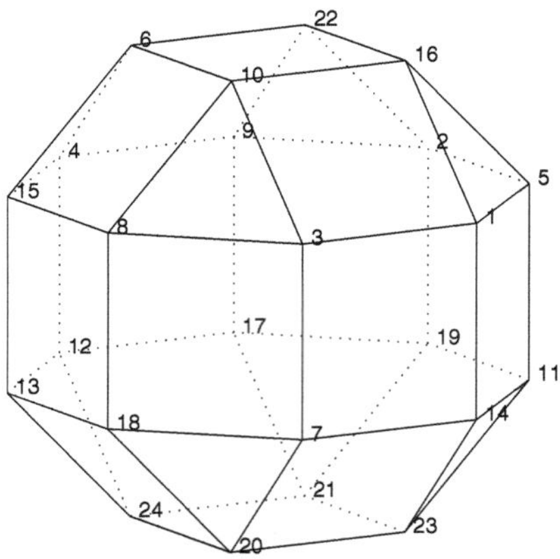

Fig. 4 The Archimedean polyhedron $\mathscr{Q}_{\mathscr{O}}$ with numbered vertexes

Assume $T = 1$ and let $\mathscr{K}$ be the cone associated to v. The loops in $\mathscr{K}$ do not wind around one axis only, therefore the action $\mathscr{A}$ restricted to $\mathscr{K}$ is coercive and we can find a minimizer u_* of $\mathscr{A}|_{\mathscr{K}}$.
From the structure of v we can restrict the minimization to the loops satisfying (6), with $M = 4$. Then by Table 1 we have

$$a_{\mathscr{R},M} \approx 1152.$$

Now consider the linear piecewise loop v defined by v, with the same velocity on each linear path. Referring to the notation of relation (7) we have $n_1 = 8$, $n_2 = 4$, so that by Table 2 we obtain

$$\mathscr{A}(v) \approx 703.2 < a_{\mathscr{R},M},$$

so that u_* does not have total collisions.

Moreover, $\mathscr{K}$ is simple and v does not wind around two axes only, therefore we can exclude also partial collisions.
By Palais' principle of symmetric criticality [15] we conclude that the minimizer u_* is actually a critical point of the unconstrained action functional.
Finally, using the theory of elliptic regularity we conclude that u_* is a smooth periodic solution of the N-body problem.

An interesting feature of this orbit is that the particles do not share a common center, i.e. relation (3) does not hold. This is a consequence of Theorem 4.2 in [8], implying that the projection of a collision-free minimizer $u_* \in \mathscr{K}$ on the unit sphere $\mathbb{S}^2$ crosses the minimum number of triangles compatible with membership to $\mathscr{K}$, i.e. the same triangles in the sequence $\mathfrak{t}$ that defines $\mathscr{K}$ (see Fig. 2, left).

In Figure 5 we sketch the motion of the $N = 24$ particles. The dashed hexahedron is used as reference for the symmetry. The bodies may be divided into 6 groups of 4 particles, each group following a choreography. The path of one of these choreographies is enhanced.

Some movies with this and similar orbits can be found at the website [9].

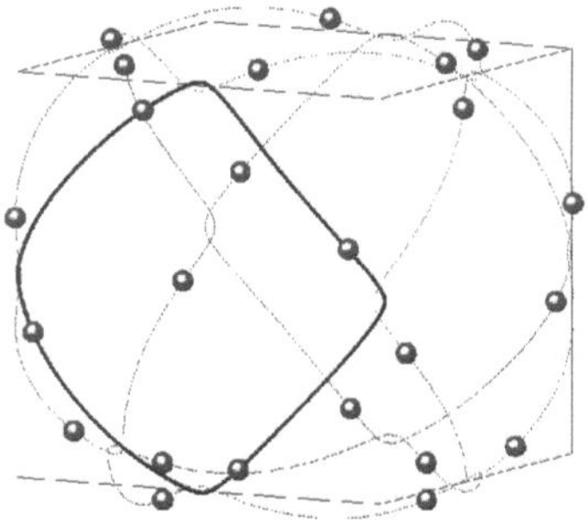

Fig. 5 Periodic motion of the $N = 24$ particles

References

1. Barutello, V., Ferrario, D.L., Terracini, S.: On the singularities of generalized solutions to n-body-type problems. Int. Math. Res. Not., Art. ID rnn 069, 1–78 (2008)
2. Chenciner, A.: Action minimizing solutions of the Newtonian n-body problem: from homology to symmetry. ICM 2002. Beijing (2002)
3. Chenciner, A.: Symmetries and "simple" solutions of the classical N-body problem. ICMP03 (2003)
4. Chenciner, A., Montgomery, R.: A remarkable periodic solution of the three-body problem in the case of equal masses., Ann. Math. **152**(2), 881–901 (2000)
5. Coti Zelati, V.: Periodic solutions for N-body type problems. Ann. Inst. H. Poincaré. Anal. Non Linéaire **7/5**, 477–492 (1990)
6. Degiovanni, M., Giannoni, F., Marino, A.: Periodic solutions of dynamical systems with Newtonian type potentials. Ann. Scuola Norm. Sup. Pisa. Cl. Sci. **15**, 467–494 (1988)
7. Ferrario, D., Terracini, S.: On the existence of collisionless equivariant minimizers for the classical n-body problem. Invent. Math. **155**, 305–362 (2004)
8. Fusco, G., Gronchi, G.F., Negrini, P.: Platonic Polyhedra, Topological Constraints and Periodic Solutions of the Classical N-Body Problem. Invent. Math. **185**, 283–332 (2011)
9. http://adams.dm.unipi.it/~gronchi/nbody.html
10. Gordon, W.B.: A minimizing property of Keplerian orbits. American Journal of Mathematics **99**(5), 961–971 (1977)
11. Kepler, J.: Mysterium Cosmographicum (1621); English version: The Secret of the Universe, translated by E. J. Aiton. Abaris Books, New York (1981)
12. Marchal, C.: How the method of minimization of action avoid singularities, Cel. Mech. Dyn. Ast. **83**, 325–353 (2002)
13. Poincaré, H .: Les méthodes nouvelles de la Mécanique Céleste (1892). Reprinted by Blanchard, Paris, Vol. **I**, p.82 (1987)
14. Moore, C.: Braids in Classical Dynamics. Phys. Rev. Lett. **70** n. 24, 3675–3679 (1993)
15. Palais, R.: The principle of symmetric criticality. Comm. Math. Phys. **69**(1), 19–30 (1979)
16. Poincaré, H.: Sur Les Solutions Périodiques et le Principe de Moindre Action. C.R.A.S, t.123, 915–918 (1896)
17. Simó, C.: New Families of Solutions in N-Body Problems, Proceedings of the third European Congress of Mathematics, Casacuberta et al. edits. Progress in Mathematics **201**, 101–115 (2001)
18. Tonelli, L.: Fondamenti del calcolo delle variazioni. Zanichelli, Vol. **I** (1921); Vol. **II** (1923)
19. Venturelli, A.: Application de la minimisation de l'action au Problème de N corps dans le plan e dans l'espace. Thesis, University Paris VII (2002)
20. Wintner, A.: The Analytical Foundations of Celestial Mechanics. Princeton University Press, New Jersey (1941)

Superprocesses as Models for Information Dissemination in the Future Internet

Laura Sacerdote, Michele Garetto, Federico Polito and Matteo Sereno

Abstract Future Internet will be composed by a tremendous number of potentially interconnected people and devices, offering a variety of services, applications and communication opportunities. In particular, short-range wireless communications, which are available on almost all portable devices, will enable the formation of the largest cloud of interconnected, smart computing devices mankind has ever dreamed about: the Proximate Internet. In this paper, we consider superprocesses, more specifically super Brownian motion, as a suitable mathematical model to analyse a basic problem of information dissemination arising in the context of Proximate Internet. The proposed model provides a promising analytical framework to both study theoretical properties related to the information dissemination process and to devise efficient and reliable simulation schemes for very large systems.

L. Sacerdote (✉)
Dipartimento di Matematica, Università degli Studi di Torino, via Carlo Alberto 10, I-10123 Torino, Italy
e-mail: laura.sacerdote@unito.it

M. Garetto
Dipartimento di Informatica, Università degli Studi di Torino, Corso Svizzera 185, I-10149 Torino, Italy
e-mail: michele.garetto@unito.it

F. Polito
Dipartimento di Matematica, Università degli Studi di Torino, via Carlo Alberto 10, I-10123 Torino, Italy
e-mail: federico.polito@unito.it

M. Sereno
Dipartimento di Informatica, Università degli Studi di Torino, Corso Svizzera 185, I-10149 Torino, Italy
e-mail: matteo@di.unito.it

A. Celletti, U. Locatelli, T. Ruggeri and E. Strickland (eds.): *Mathematical Models and Methods for Planet Earth.* Springer INdAM Series 6, DOI 10.1007/978-3-319-02657-2_13,

1 Introduction

So many events had a significant impact on our life in these last 50 years that it can be difficult to narrow it down to a few. Flights have restricted our perception of distances, radio and television have made people aware of facts and news, dangerous illnesses have disappeared thanks to vaccines while new antibiotics have helped the recovery from many dangerous diseases, medicine has decreased childhood mortality in many countries. The list could continue but we focus here on one of the most sudden events of our last thirty years that impacts our daily life everywhere: the advent of the Internet. More than one billion people on Earth use the Internet nowadays.

Due to the importance of the Internet for our planet development, in Section 2 we briefly list some important facts about it and its diffusion. Then in the same section we present the past and the future role that wireless technology played and will play for the development of possible future services of the Internet.

Next, we move to the problem considered in this paper, i.e., the dissemination of information in large, disconnected mobile ad-hoc networks [28]. Such networks will arise naturally in a possible Future Internet scenario, where an increasing number of users carrying existing and novel wireless devices (smartphones, tablets, laptops, smartwatches, smartglasses, etc.) will form the so-called *Proximate Internet*, i.e., the largest cloud of interconnected, smart computing devices mankind has ever dreamed about. In the *Proximate Internet*, users will be able to directly communicate among themselves exploiting short-range radio communications, enabling a variety of interesting applications, and making the search and distribution of information much more efficient (in terms of spectrum usage, energy and monetary costs for the users) than traditional (cellular) networks.

In this scenario, information spreads among the wireless nodes in an epidemic fashion, i.e., like an infection in a human population. Hence mathematical models developed for epidemiology can be applied, after properly adapting them to our specific context. Indeed, users carrying wireless devices are typically in motion over a certain area (e.g., a city), and they can communicate with other devices within a limited communication range. Hence the epidemic model must account for these fundamental features.

The availability of mathematical models is of paramount importance to understand and design future applications running in the Proximate Internet. For example, models can predict the information dissemination speed, and thus the delays incurred to reach far-away users, and the efficiency of the dissemination in terms of area coverage, as function of a variety of system parameters.

The rest of the paper is organized as follows. In Section 2 we start with a general introduction to the topic of mathematical modelling of the Internet. In Section 3 we provide a high-level description of the specific scenario of wireless content distribution that we consider in our work, before going into technical details. In Section 4, we introduce some necessary background to the mathematical tools adopted in our model. In particular, we briefly recall how to relate the discrete space and time branching process, i.e. the Galton–Watson model, with its continuous time counterpart, i.e. the continuous space branching process. Then we add to each particle of the branching model a movement mechanism and we introduce the resulting super Brownian motion as a measure valued process suitable to describe the information dissemination in a wireless cloud. For each of these processes we recall those properties which are most significant for our modeling purposes.

In Section 5 we present our model as an instance of super Brownian motion, explaining why this representation provides a much powerful performance evaluation methodology than traditional detailed simulations. We conclude with directions of future research and a with a detailed reference section which can be a helpful tool for the interested reader.

2 Internet, Planet Earth and Mathematics

It is impressive to realize that the Internet has existed for a so short period cosidering how strongly it changed the life on our Planet and the speed of its penetration worldwide. In Figures 1 and 2 we illustrate the growth and the current presence of the Internet in Europe. The Internet is also a formidable tool for the evolution of underdeveloped countries where it allows the diffusion of books and culture as well as to give long distance medical support to people living in unreachable areas of our planet. The list of the changes determined in everyday life is incredibly long and

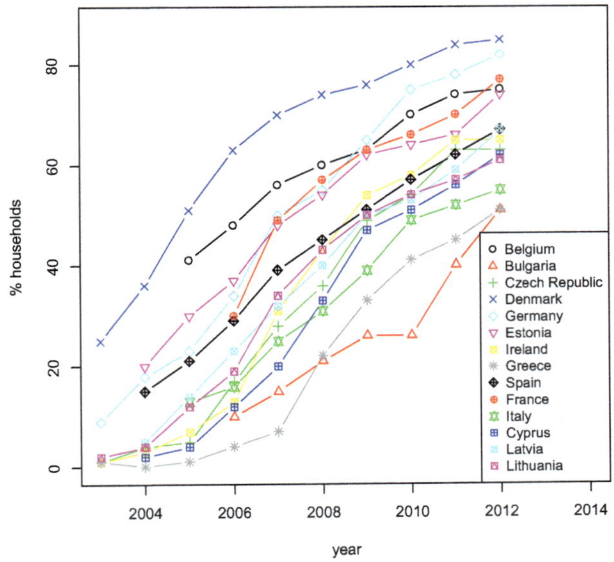

Fig. 1 Percentage of households with broadband access with at least one member aged 16 to 74. Source: Eurostat, the statistical office of the European Union (http://ec.europa.eu/eurostat); see also Fig. 2

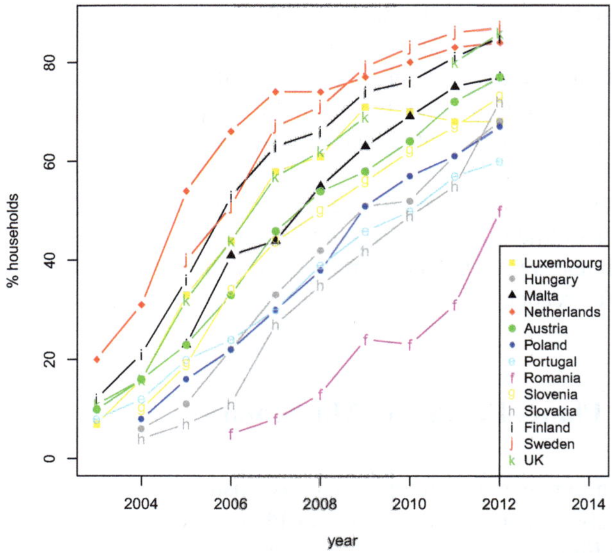

Fig. 2 Percentage of households with broadband access with at least one member aged 16 to 74. Source: Eurostat, the statistical office of the European Union (http://ec.europa.eu/eurostat); see also Fig. 1

this short contribution is not the place where to discuss such important political and social events.

We prefer to just provide an example taken from our everyday experience as mathematical researchers, that clearly shows the fundamental role played by the Internet. The oldest between us still have memory of the incredible waste of time related with any bibliographic research less then twenty years ago. Youngest have never spent hours in a department library consulting books of Mathematical Reviews, writing down the references and then moving to look for the suggested journals, climbing ladders and moving heavy volumes up and down. Now we simply Google the title or the subject of the paper eventually using Mathematical Reviews online. When we recognize the title of an interesting paper a further click miraculously let it appear on our screen. Going further, who has never checked the definition of some mathematical object on Wikipedia for a fast suggestion?

The evolution of the Internet is so fast that we get immediately used to any change and we finally disregard principia allowing the creation of these new tools. However the improvements of the Internet are often related with mathematical results since its first appearance. To cite a famous example we recall that the first network between computers was made possible by Kleinrock with his mathematical theory of packet networks, the technology underlying today's Internet [30]. A more recent example is the mathematical work of [8] underlying the algorithms of Google search engine (many mathematicians are still working on their improvements).

The Internet is also a source of new questions of high mathematical interest. To give a couple of examples, let us think to the preferential attachment phenomenon. In the Internet, preferential attachment makes nodes (that can represent either routers, or web pages, or people in social networks) with higher degree to have stronger ability to grab links added to the network. It is well known that preferential attachment generates scale-free networks, i.e. networks whose degree distribution follows a power law, at least asymptotically [4, 11, 21]. Various analytical results are known for this model but, for example, its clustering coefficient is only numerically known and its analytical expression is still an open problem.

Many other examples of open mathematical problems could be cited, determined for example by congestion problems and transmission optimization [49].

We can pursue different philosophies to deal with fundamental problems related to the current and future Internet which translates into different needs for mathematical models and analyses. Next generation Internet involves different topics that can be roughly classified under the primary networking function to which they are related: routing, management and control (centralized or scalable), security (additional overlap or part of the architecture), challenged network environments (continuously connected–intermittent connectivity; tools for connectivity), content delivery (robust and scalable methods) are a possible classification of the involved topics.

Furthermore, each of these subjects can be studied both under the so called *clean-slate* paradigm or as corrections and improvements of present situation. The *clean-slate* design is the philosophy adopted by a project started at Stanford University[1], based on the belief that the current Internet has significant deficiencies that need to be solved before trying to improve future features of our global communication infrastructure. The basic idea is to explore what kind of Internet we would design if we were to start from scratch, with 20–30 years of hindsight. In this framework many interesting mathematical problems arise. A good example coming from the area of network security are the so called *adaptive resilient hosts*, based on the general idea that a system should adapt to attacks by changing process code, perhaps also limiting services temporarily[2]. Other typical examples arise in the context of congestion control algorithms [1]. On the other hand, many studies renounce the clean-slate approach, and just consider improvements of existing protocols and codes.

Almost invariably, while modelling the Internet, evident difficulties arise both technically and mathematically, when the focus moves from an isolated problem to the whole Internet, which is an incredibly complex system whose evolution is not simply driven by technological factors.

In our work we focus on a specific problem related to the transmission of files to large populations of users. Hence we recall some basic facts about this topic. Wireless and wireline digital infrastructure currently coexist to transmit digital and analog files of data, voice, video, text, image, fax and streaming media. This coexistence will grow up in the next future giving rise to new Internet services, day after day more user-friendly. Due to the growing importance of the wireless technology, we provide here a brief summary of its history. Wireless transmission has its ori-

---

[1] Clean Slate Program, Stanford University, http://cleanslate.stanford.edu.

[2] CRASH – Clean-slate Resilient Adaptive Secure Hosts, Cornell University, http://www.nuprl.org/crash/introextended.php.

gins around 1895 when Tesla and Marconi independently developed the wireless telegraphy but its mathematical foundations date back to 1864 when J.C. Maxwell mathematically predicted the existence of radio waves. Mathematical results were then experimentally verified by D.E. Hughes who first transmitted Morse code. After wireless telegraph in 1900 wireless technology was used to transmit voice over the first radio and following developments determined the diffusion of radios around the globe. Then wireless technology did not give rise to new epochal instruments until the digital age of the Internet in the latter part of the twentieth century, with the appearance of digital cellular phones, mobile networks, wireless network access and so on. However the beginning of the Internet was wireline based and only recently the power of wireless communications has been unleashed.

L. Kleinrock in his paper on the History of Internet and its Flexible Future [31] recognizes the following five phases through which the Internet will evolve in the next few years: nomadic computing, smart spaces and smart networks, ubiquitous computing, platform convergence and intelligent agents. Indeed there is an increasing number of travelling users that requests trouble free Internet services from any device, any place and any time (nomadic computing). Meanwhile it is no more science fiction to think to a cyberspace where intelligent sensors give the alarm in the presence of sudden health problems of aged people or where we can use voice commands to interact with devices in a room (smart spaces and smart networks). Ubiquitous computing is related with the necessity to have Internet services available wherever, for example to consult maps or train timetables. As Kleinrock remarks it is ridiculous to travel with a number of electronic devices: cellular phones, notebook computers, clocks, microphones, cameras, batteries, chargers. All these devices should converge in one, equipped with all necessary technologies. Finally, intelligent agents, i.e., autonomous software modules acting on data, observing trends, carrying out tasks or adapting to the environment, will generate an increasing amount of traffic that should use wireless channels.

Nowadays wireless technology is becoming absolutely fundamental both to access the Internet in the traditional sense and to communicate locally with nearby users and objects, like sensors and actuators (i.e., the Internet of Things). This evolution is made possible by the increasing popularity of portable devices, and the availability of a tremendous number of different services and applications enabled by these devices.

3 Wireless information dissemination: system overview

In our problem, we exploit device-to-device (wireless) communications to distribute information from one or more sources to a large number of mobile users belonging to the same *cloud* (the Proximate Internet). The exploitation of local, short-range communications (using for example WiFi or Bluetooth technology) permits to offload cellular networks (3G/4G), with significant advantages both for cellular operators and for the users, who can obtain almost for free the contents they are interested in directly from other nearby users.

More specifically, let us focus on a piece of information (an entire file or just a portion of it), hereinafter called the message, which is initially stored at one or more nodes (called the seeds). These source nodes might have retrieved the message from the Internet by other means (i.e., different wired or wireless technologies) or they can have generated it locally. The message can be transmitted to other nodes interested in it, but only within a limited communication range, depending on the specific wireless technology (in the order or a few tens of meters). By so doing, the message gets replicated in the wireless cloud in an epidemic fashion, reaching more and more nodes with the passing of time. At the same time, nodes who have already received the message may stop contributing to the dissemination process, due to several reasons (the application is closed, the wireless interface is switched off, the message is cancelled from the local memory to make room to other messages, etc). Furthermore, users can move while carrying their devices, and we assume that the movements are random and independent from user to user.

Our goal is to define a suitable mathematical framework to model the above message replication process in the case of a very large number of users, so that we can study the dynamics of wireless information dissemination over a large area (like a big city). The model should be able to estimate basic performance metrics such as:

1. The delay associated to the transfer of the information, as function of the distance from the source.
2. An index of the achieved city-level coverage, and in particular a measure of the possible zones that will never be reached by the information.
3. The probability that an uncovered zone will be covered again after a reasonable delay.

The availability of such a model would be very useful to plan and design applications exploiting the proximate Internet. For example, it could be used to optimize the number and locations of the initial seeds, trading off the various costs associated to bandwidth/energy/memory constraints, according to different objective functions.

4 Mathematical background: super Brownian motion

In this section we present a brief description of the simplest superprocess that can be defined, arising as a weak limit of standard branching Brownian motions, the so-called super Brownian motion. In the most basic model of super Brownian motion we have critical behaviour, branching is quadratic, and jumps are not admitted. Plainly, more general models of super Brownian motions can be constructed, as we will see in the following. We start by analyzing the underlying branching structure and then we will add movement to the model.

Let us therefore first consider separately the branching structure. In practice the evolution of the number of nodes (i.e., the amount of messages disseminated in the network) is governed by continuous state branching processes (CSBPs), introduced by [24] and studied by various researchers during the past decades (see for example [6, 10, 18, 22, 25, 27, 34–36, 38, 40, 42, 48]). In order to understand what a CSBP is, we start by describing a related discrete time and discrete state space model: the so-called Galton–Watson process. In general a Galton–Watson process [2, 3, 23, 52] models the evolution of a population composed by individuals acting independently and behaving homogeneously. Individuals live for only one generation (one time step) then they die producing a random number Z of offsprings with a non negative discrete probability distribution $\mathbb{P}(Z=z)$, $z \in \mathbb{N} \cup \{0\}$ and with mean value $\mathbb{E}Z = \zeta < \infty$. Let us now write N_m, $m \in \mathbb{N} \cup \{0\}$ for the process counting the number of individuals in the population at time m. We put $N_0 = 1$ for simplicity (a single initial progenitor) and write that

$$N_{m+1} = \sum_{j=1}^{N_m} Z_j, \qquad m \in \mathbb{N} \cup \{0\}, \tag{1}$$

where Z_j, are independent and identically distributed random variables representing the random number of offsprings generated by individual j. By means of simple calculations it can be proven that the mean value $\mathbb{E}N_m = \zeta^m$. The Galton–Watson process is critical, subcritical or supercritical if $\zeta = 1$, $\zeta \in (0,1)$ or $\zeta > 1$, respectively. See Figure 3 for an example of a realization of a Galton–Watson counting process together with the associated underlying tree coding the whole genealogy. For more information on Galton–Watson and related processes the reader can refer to the already cited classical o more modern references.

Consider now a Markov process Y_t^x, $t \geq 0$, taking values in the positive real line, starting from point $x \in \mathbb{R} \cup \{0\}$ and having right-continuous with left limits (càdlàg) paths. We refer to [47] or [7] for an introduction to the general theory of Markov processes. We say that Y_t^x satisfies the branching property if

$$Q_t(x+z,\cdot) = Q_t(x,\cdot) * Q_t(z,\cdot), \tag{2}$$

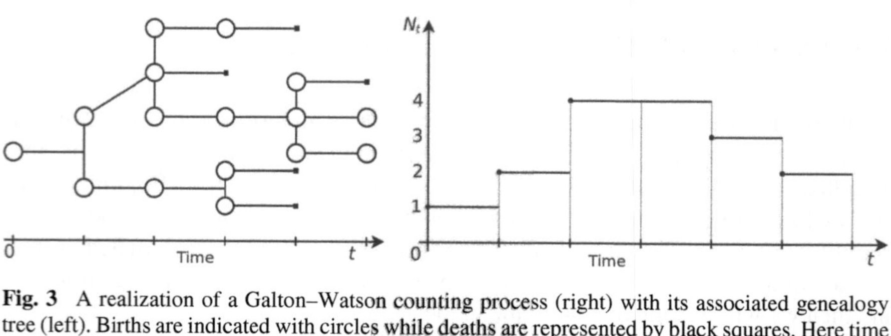

Fig. 3 A realization of a Galton–Watson counting process (right) with its associated genealogy tree (left). Births are indicated with circles while deaths are represented by black squares. Here time is continuous but $m = [t]$. In this figure it is possible to appreciate how a complete tree is far more rich in information than its related counting process

where $*$ represents the convolution operator and $(Q_t)_{t\geq 0}$ is the associated transition semigroup. This can be defined by its Laplace transform as

$$\mathbb{E}\exp(-\mu Y_t^x) = \int_0^\infty \exp(-\mu y)\, Q_t(x,\mathrm{d}y) = \exp(-x v_t(\mu)), \qquad \mu > 0, \tag{3}$$

where $x \in \mathbb{R}^+ \cup \{0\}$ is the starting point (initial population density) and the mapping $t \mapsto v_t(\mu)$ is the unique positive solution of the integral equation

$$v_t(\mu) = \mu - \int_0^t \phi(v_s(\mu))\,\mathrm{d}s, \qquad t \geq 0. \tag{4}$$

The function ϕ in the previous equation is called the branching mechanism of the continuous state branching process and in general it can be of the form [48]

$$\phi(z) = bz + cz^2 + \int_0^\infty (e^{-zy} - 1 + zy)\,\pi(\mathrm{d}y), \qquad z \geq 0, \tag{5}$$

where $(y \wedge y^2)\pi(\mathrm{d}y)$ is a finite measure concentrated on the positive real line and $b \in \mathbb{R}$, $c \in \mathbb{R}^+ \cup \{0\}$. Note that the above representation is related to the Lévy–Khintchine representation for the characteristic function of a Lévy process [5,32,46]. The three different cases of $b > 0$, $b < 0$, and $b = 0$ correspond respectively to the supercritical, subcritical, and critical cases. As we recalled above one of the simplest possible branching mechanisms is the quadratic (or binary) branching which corresponds to

$$\phi(z) = cz^2, \qquad c \in \mathbb{R}^+. \tag{6}$$

Continuous state branching processes with quadratic branching are of the critical type and are known in literature as the Feller's branching diffusions [18,44]. From (3) it is possible to determine the mean behaviour

$$\mathbb{E}Y_t^x = \int_0^\infty y\, Q_t(x,\mathrm{d}y) = x\exp(-bt), \qquad t \geq 0, x \in \mathbb{R}^+ \cup \{0\}, \tag{7}$$

for the general case (5). When the branching mechanism reduces to (6) we obviously obtain that $\mathbb{E}Y_t^x = x$ which characterizes a critical behaviour. In turn, when the branching is quadratic it is immediate to calculate an explicit form of the function $v_t(\mu)$ (see for example [36]) as

$$v_t(\mu) = \frac{\mu}{1 + ct\mu}. \tag{8}$$

Continuous state branching processes are very well studied models of population evolution. For more in-depth results standard references include [6,24,32,34–36,48, 51], [39, Chap. 3], and many others. Furthermore, it is worthy of notice that CSBPs arise naturally as weak limits of a sequences of rescaled Galton–Watson processes when waiting times between splits become negligible (see for example [23,35,36]). This last consideration will prove very useful in the interpretation of super Brownian motion.

Consider therefore the d-dimensional Euclidean space $\mathbb{R}^d$ and a random measure (see [26]) on it determining the location of particles undergoing branching (both births and deaths) at time t, written as

$$X_t = \frac{1}{\beta} \sum_j \delta_{B_t^j}, \qquad t \geq 0. \tag{9}$$

In (9), for each alive particle j, the process $B_t^j, t \geq 0$, is a d-dimensional Brownian motion on $\mathbb{R}^d$ with initial starting point $x_0^j \equiv B_0^j$. Furthermore B_t^j is independent of $B_t^i, i \neq j$, $\delta_{B_t^j}$ is a Dirac delta measure on B_t^j, and β is a normalizing factor related to the population size. The process (9) is a measure valued process on $\mathbb{R}^d$ representing simultaneous motion of non-interacting particles.

Beside this, each particle undergoes branching such that the total number of alive particles follows a Galton–Watson process:

$$W_t = \frac{1}{\beta} N_{[t]}, \qquad t \geq 0. \tag{10}$$

Note that branching occurs at time $1, 2, 3, \ldots$, that is the waiting times between branching events are deterministic and all equal to unity. See in Figure 4 a sketch of a possible realization of the complete process where particles move following independent Brownian motions and branch at fixed times $t_1 = 1$, $t_2 = 2$, and so forth.

We are in fact interested in the weak limit of the rescaled processes

$$X_t^k = \frac{1}{\beta_k} \sum_j \delta_{B_t^{j,k}}, \qquad t \geq 0, k \geq 1, \tag{11}$$

with $B_0^{j,k} \equiv x_0^{j,k}$. Clearly, aside the sequence $(X_t^k)_{k \geq 1}$, we consider the related sequence of processes

$$W_t^k = \frac{1}{\beta_k} N_{[kt]}^k, \qquad t \geq 0, k \geq 1, \tag{12}$$

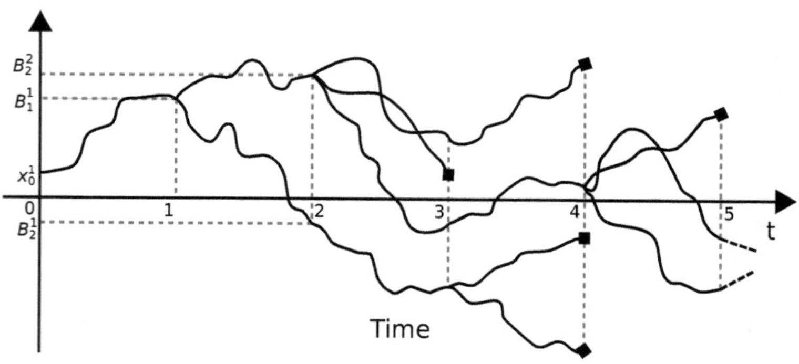

Fig. 4 A sketch of a possible realization of the genealogy tree with Brownian movement. Death of particles is represented by black squares

which measures the total mass present (amount of messages in the network) and where the branching waiting times are all equal to t/k. The relation which links the two processes W_t^k and X_t^k is clearly that $\langle X_t^k, 1\rangle = W_t^k$. Let us now consider the space $M(\mathbb{R}^d)$ of finite measures on $\mathbb{R}^d$. If the sequence of the initial measures converges towards the finite measure $\xi \in M(\mathbb{R}^d)$ we have that $(X_t^k, t \geq 0)_{k\geq 1}$ converges weakly towards an $M(\mathbb{R}^d)$-valued Markov process $X_t, t \geq 0$ (see [41], Theorem 2.1.9, [17], Chap. 9). Correspondingly $(W_t^k, t \geq 0)_{k\geq 1}$ converges weakly to a continuous state branching process with a general branching mechanism (5). Since the underlying spatial motion is of Brownian type, the process obtained is called *ϕ-super Brownian motion*. In the simplest case of a quadratic (or binary) branching it is simply called *super Brownian motion*. Relevant references for super Brownian motion and superprocesses in general are [12, 16, 19, 33, 36, 37, 39, 43, 45]. Moreover, it should be noted that also more general superprocess are already described in the literature (see e.g. [13–15, 20, 50]).

5 Wireless information dissemination: basic model and future directions

Now we can explain how the information dissemination scenario introduced in Section 3 can be represented in terms of a super Brownian motion of the simplest kind, as described in Section 4.

To do so, some simplifying hypotheses are necessary in order to obtain a tractable model of the considered system. We assume that each user independently moves over the infinite plane according to a simple random walk. We know that realistic models of human mobility suggest that individuals' movements should be modeled through Lévy flights, i.e. as a random walk in which the step-lengths have a probability distribution that is heavy-tailed [9]. Alternatively truncated Levy flights were proposed after analysing data [21]. However, for a first simplified model we consider the movements of each user as a simple random walk, allowing us to adopt a standard Brownian motion to describe the macroscopic mobility of each node.

While moving, nodes come in contact with other nodes (i.e., other nodes fall in the communication range of the considered node), allowing the opportunistic, direct transfer of the message of interest. A standard computation, that takes jointly into account the node density, the transmission range, and the node speed, permits computing the rate λ at which the message gets duplicated. We do not repeat the details of this computation, which can be found in [29]. Hence, the birth rate λ can be considered as a primitive system parameter.

We instead denote by μ the death rate of each node (i.e., the rate at which a node transits to a state in which it no longer contributes to the message dissemination), which is supposed to be given. At last, let σ^2 be the infinitesimal variance of the Brownian motion describing the macroscopic mobility of each node. Moreover, we assume that, at time $t = 0$, there is a given number N_0 of seeds, initially co-located at the origin of the plane.

Disregarding for now the nodes' movements, these hypotheses characterize the previously presented *Galton–Watson tree*, i.e. the model that describes the evolution of a population that starts with a given size at time $t = 0$. Depending on the value of λ with respect to μ the population becomes extinct with probability 1 (subcritical case $\lambda < \mu$ and critical case $\lambda = \mu$) or has a positive probability to survive forever (supercritical case: $\lambda > \mu$). When the population becomes sufficiently large its size can be modelled as a *continuous state branching process*, as we pointed out in the previous paragraph. Hence the number of nodes storing the message of interest can be described through these processes.

However, nodes also move according to a random walk, and we are especially interested in characterizing the distribution on the plane of the nodes who are currently storing the message, as function of time. In our brief review in Section 4 we pointed out that, when the size of the population is large and microscopic displacements of the nodes are frequent enough, one can rescale the Galton–Watson tree with moving particles obtaining a *super Brownian motion*.

For the super Brownian motion, results are available for the asymptotic speed of diffusion. These results can be directly used to study the delay incurred by the message delivery to far-away users.

In our case nodes move over $\mathbb{R}^2$ and the super Brownian motion is known to have singular distribution on $\mathbb{R}^2$. Furthermore this distribution is uniform on its random support. Then the study of the support properties of a super Brownian motion is a possible approach to deal with the city coverage problem: we assume that each device covers a region of radius r, equal to its transmission range. This leads to study the coverage of the support of the Super Brownian Motion through balls. The aim of this study should be to determine which percentage of the plane is covered by at least one ball. Possible zones that will never be reached by the information could also be investigated following this approach.

Beside the above theoretical results, our model also allows us to devise efficient simulation schemes for scenarios in which no analytical results are available, such as those in which the birth and/or death rates of the nodes, and/or the infinitesimal variance of the associated brownian motion, are a function of space and/or time. On this regard, we emphasize that a brute-force simulation approach in which each node is modelled in all details becomes infeasible for large number of nodes. In particular, in the super-critical regime it is basically impossible to simulate the system after a very short time, due to the exponential growth of the number of nodes. Therefore, alternate simulation approaches are necessary.

Simulations represent the only methodology to study scenarios of particular interest, in which the parameters of the process are not homogeneous in space (presence of lakes, zones without inhabitants and so forth). The available techniques for these simulations do not consider the singular nature of the super Brownian motion. Hence reliable and efficient simulation methods will be the subject of future work, together with the study of more theoretical results of interest for our application scenario.

Acknowledgements The authors have been supported by project AMALFI (Università di Torino/ Compagnia di San Paolo).

References

1. Alizadeh, M., Javanmard, A., Prabhakar, B.: Analysis of DCTCP: stability, convergence, and fairness. In: Proceedings of the ACM SIGMETRICS joint international conference on Measurement and modeling of computer systems, pp. 73–84 (2011)
2. Athreya, K.B., Jagers: Classical and modern branching processes. Springer, New York (1997)
3. Athreya, K.B., Ney, P.E.: Branching processes, vol. 28. Springer, New York (1972)
4. Barabási, A-L., Albert, R.: Emergence of scaling in random networks. Science **286**(5439), 509–512 (1999)
5. Bertoin, R.: Lévy processes, vol. **121**. Cambridge University Press, Cambridge (1998)
6. Bingham, N.H.: Continuous branching processes and spectral positivity. Stochastic Processes and their Applications **4**(3), 217–242 (1976)
7. Blumenthal, Getoor, R.K.: Markov processes and potential theory. Courier Dover Publications, New York (1968)
8. Brin, S., Page, L.: The anatomy of a large-scale hypertextual Web search engine. Computer networks and ISDN systems **30**(1), 107–117 (1998)
9. Brockmann, D., Hufnagel, L., Geisel, T.: The scaling laws of human travel. Nature **439**(7075), 462–465 (2006)
10. Caballero, M.E., Lambert, A., Bravo, G.U.: Proof(s) of the Lamperti representation of continuous-state branching processes. Probability Surveys **6**, 62–89 (2009)
11. Candia, J., González, M.C., Wang, P., Schoenharl, T., Madey, G., Barabási, A-L.: Uncovering individual and collective human dynamics from mobile phone records. Journal of Physics A, **41**(22), 224015 (2008)
12. Dawson, D.A.: Measure-valued Markov processes. In École d'été de probabilités de Saint-Flour XXI-1991, pp. 1–260. Springer-Verlag, Berlin Heidelberg (1993)
13. Dawson, D.A., Li, Z., Wang, H.: Superprocesses with dependent spatial motion and general branching densities. Electronic Journal of Probability **6**, 1–33 (2001)
14. Engländer, J., Fleischmann, K.: Extinction properties of super-Brownian motions with additional spatially dependent mass production. Stochastic processes and their applications **88**(1), 37–58 (2000)
15. Engländer, J., Kyprianou, A.E.: Local extinction versus local exponential growth for spatial branching processes. The Annals of Probability **32**(1A),78–99 (2004)
16. Etheridge, A.: An introduction to superprocesses, vol. **20**. American Mathematical Society (2000)
17. Ethier, S.N., Kurtz, T.G.: Markov processes: characterization and convergence, vol. 282. Wiley, New Jersey (2009)
18. Feller, W.: Diffusion processes in genetics. In Proc. Second Berkeley Symp. Math. Statist. Prob, vol. **227**, p. 246 (1951)
19. Fitzsimmons, P.J.: Construction and regularity of measure-valued Markov branching processes. Israel Journal of Mathematics **64**(3), 337–361 (1988)
20. Fleischmann, K., Klenke, A.: The biodiversity of catalytic super-Brownian motion. Annals of Applied Probability, pp. 1121–1136 (2000)
21. Gonzalez, M.C., Hidalgo, C.A., Barabasi, A-L.: Understanding individual human mobility patterns. Nature **453**(7196), 779–782 (2008)
22. Grey, D.R.: Asymptotic behaviour of continuous time, continuous state-space branching processes. Journal of Applied Probability, pp. 669–677 (1974)
23. Haccou, P., Jagers, P., Vatutin, V.A.: Branching processes: variation, growth, and extinction of populations, vol. **5**. Cambridge University Press, Cambridge (2005)
24. Jiřina, M.: Stochastic branching processes with continuous state space. Czechoslovak Mathematical Journal **8**(2), 292–313 (1958)
25. Kale, M., Deshmukh, S.R.: Estimation in a continuous state space branching process. Biometrical journal **34**(8), 1001–1006 (1992)
26. Kallenberg, O.: Random measures. Akademie-Verlag, Berlin (1976)

27. Kashikar, A.S., Deshmukh, S.R.: Second order branching process with continuous state space. Statistics & Probability Letters (2012)
28. Klein, D.J., Hespanha, J., Madhow, U.: A reaction-diffusion model for epidemic routing in sparsely connected MANETs. In: INFOCOM, 2010a Proceedings IEEE, pp. 1–9 (2010)
29. Klein, D.J., Hespanha, J., Madhow, U.: A Reaction-diffusion Model For Epidemic Routing In Sparsely Connected MANETs. In: Proceedings of IEEE INFOCOM 2010b, pp. 884–892 (2010)
30. Kleinrock, L.: Information Flow in Large Communication Nets. Phd thesis proposal, Massachusetts Institute of Technology (1961)
31. Kleinrock, L.: History of the Internet and its flexible future. Wireless Communications. IEEE **15**(1), 8–18 (2008)
32. Kyprianou, A.E.: Introductory lectures on fluctuations of Lévy processes with applications. Springer-Verlag, Berlin Heidelberg (2007)
33. Kyprianou, A.E., Liu, R-L., Murillo-Salas, A., Ren, Y-X.: Supercritical super-Brownian motion with a general branching mechanism and travelling waves. Annales de l'Institut Henri Poincaré-Probabilités et Statistiques **48**(3), 661–687 (2012)
34. Lamperti, J.: Continuous-state branching processes. Bull. Amer. Math. Soc, **73**(3), 382–386 (1967)
35. Lamperti, J.: The limit of a sequence of branching processes. Probability Theory and Related Fields **7**(4), 271–288 (1967)
36. Le Gall, J-F.: Spatial branching processes, random snakes and partial differential equations. Birkhäuser-Verlag, Basel (1999)
37. Le Gall, J-F., Perkins, E.: The Hausdorff measure of the support of two-dimensional super-Brownian motion. The Annals of Probability, pp. 1719–1747 (1995)
38. Li, Y.: A weak limit theorem for generalized Jiřina processes. Journal of Applied Probability, pp. 453–462 (2009)
39. Li, Z.: Measure-valued branching Markov processes. Springer-Verlag, Berlin Heidelberg (2011)
40. Li, Z.: Path-valued branching processes and nonlocal branching superprocesses. arXiv preprint arXiv:1203.6150 (2012)
41. Li, Z.: Continuous-state branching processes. arXiv:1202.3223 [math.PR] (2012)
42. Li, Z-H.: Branching processes with immigration and related topics. Frontiers of Mathematics in China **1**(1), 73–97 (2006)
43. Mörters, P.: The average density of super-Brownian motion. Annales de l'Institut Henri Poincare (B) Probability and Statistics **37**(1), 71–100 (2001)
44. Pardoux, E., Wakolbinger, A.: From Brownian motion with a local time drift to Feller's branching diffusion with logistic growth. Electronic Communications in Probability **16**, 720–731 (2011)
45. Perkins, E.: Dawson-watanabe superprocesses and measure-valued diffusions. Lectures on probability theory and statistics, pp. 125–329 (2002)
46. Sato, K.: Lévy processes and infinitely divisible distributions. Cambridge University Press, Cambridge, UK (1999)
47. Sharpe, M.: General theory of Markov processes, vol. 133. Academic Press, Boston, MA (1988)
48. Silverstein, M.L.: Continuous state branching semigroups. Zeitschrift für Wahrscheinlichkeitstheorie und Verwandte Gebiete **14**(2), 96–112 (1969)
49. Srikant, R.: The mathematics of Internet congestion control. Birkhäuser-Verlag, Boston (2004)
50. Wang, H.: State classification for a class of interacting superprocesses with location dependent branching. Elect. Commun. Probab **7**(16), 157–167 (2002)
51. Watanabe, S.: On two dimensional Markov processes with branching property. Transactions of the American Mathematical Society **136**, 447–466 (1969)
52. Watson, H.W., Galton, F.: On the probability of the extinction of families. The Journal of the Anthropological Institute of Great Britain and Ireland **4**, 138–144 (1875)

Appendix
Pictures from INdAM Workshop

The pictures of the participants included in this section are by Franco Forci

Christiane Rousseau (Université de Montréal)

Alessandra Celletti (Università di Roma Tor Vergata)

Laure Saint-Raymond (Università di Parigi 6, ENS Paris)

From left to right: Ugo Locatelli (Università di Roma Tor Vergata) Alessandra Celletti (Università di Roma Tor Vergata) Elisabetta Strickland (Università di Roma Tor Vergata) Tommaso Ruggeri (Università di Bologna)

Maria Letizia Bertotti (Libera Università di Bolzano)

Paolo Dai Prà (Università di Padova)

Andrea Tosin (IAC, CNR, Roma)

Anne Lemaitre (FUNDP, Università di Namur, Belgio)

Luigi Preziosi (Politecnico di Torino)

Ettore Perozzi (Deimos Space, Roma)

Gerard Gomez (Università di Barcellona)

Adriano Barra (Università La Sapienza, Roma)

Jacques Laskar (IMCCE, Parigi)

Mirko Degli Esposti (Università di Bologna)

Antonello Provenzale (ISAC-CNR, Torino)

Giovanni Federico Gronchi (Università di Pisa)

Laura Sacerdote (Università di Torino)

Participants

MIX
Papier aus verantwortungsvollen Quellen
Paper from responsible sources
FSC® C105338

If you have any concerns about our products,
you can contact us on
ProductSafety@springernature.com

In case Publisher is established outside the EU,
the EU authorized representative is:
Springer Nature Customer Service Center GmbH
Europaplatz 3, 69115 Heidelberg, Germany

Printed by Libri Plureos GmbH
in Hamburg, Germany